THE WORKS

ANATOMY OF

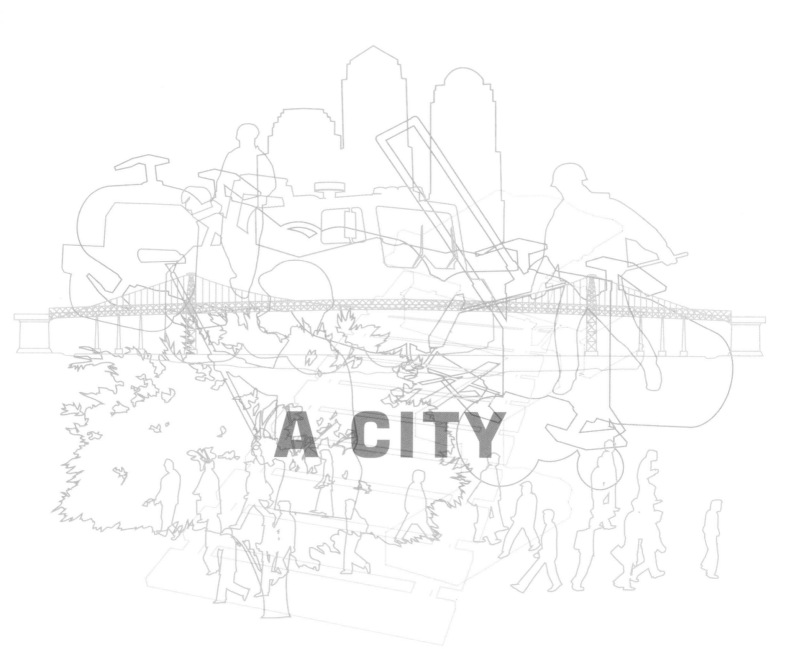

A CITY

Kate Ascher

Researched by Wendy Marech

Designed by Alexander Isley Inc.

PENGUIN BOOKS

PENGUIN BOOKS

Published by the Penguin Group

Penguin Group (USA) Inc., 375 Hudson Street, New York, New York 10014, U.S.A.

Penguin Group (Canada), 90 Eglinton Avenue East, Suite 700, Toronto,
Ontario, Canada M4P 2Y3 (a division of Pearson Penguin Canada Inc.)

Penguin Books Ltd, 80 Strand, London WC2R 0RL, England

Penguin Ireland, 25 St Stephen's Green, Dublin 2, Ireland (a division of Penguin Books Ltd)

Penguin Group (Australia), 250 Camberwell Road, Camberwell,
Victoria 3124, Australia (a division of Pearson Australia Group Pty Ltd)

Penguin Books India Pvt Ltd, 11 Community Centre,
Panchsheel Park, New Delhi – 110 017, India

Penguin Group (NZ), 67 Apollo Drive, Rosedale, North Shore 0632,
New Zealand (a division of Pearson New Zealand Ltd)

Penguin Books (South Africa) (Pty) Ltd, 24 Sturdee Avenue,
Rosebank, Johannesburg 2196, South Africa

Penguin Books Ltd, Registered Offices:
80 Strand, London WC2R 0RL, England

First published in the United States of America by The Penguin Press,
a member of Penguin Group (USA) Inc. 2005
Published in Penguin Books 2007

1 3 5 7 9 10 8 6 4 2

ISBN 1-59420-071-8 (hc.)
ISBN 978-0-14-311270-9 (pbk.)
CIP data available

Printed in the United States of America
Set in FF Eureka and Square 721
Designed by Alexander Isley Inc.

through their pipes—while millions more are carried away as waste. All the while, vast amounts of power are consumed by their homes and businesses and millions of gigabytes of data flow through their telecom wires.

Rarely does a resident of any of the world's great metropolitan areas pause to consider the complexity of urban life or the myriad systems that operate round the clock to support it. He or she wakes up in the morning to turn on a tap, switch on a light, flush a toilet or perhaps grab a banana—little knowing how much effort, on the part of how many people, goes into making the simplest morning routine possible. The rest of the day is also deceptively simple: crossing a street, riding the subway, taking out the garbage—even the most mundane domestic tasks would be impossible without the far-reaching, complex, and often invisible network of infrastructure that supports them.

While this holds true for urban life across the globe, no city is more dependent on its infrastructure than New York. A vertical as well as a horizontal city, power is king: without it, the two things that move more bodies than any other—the subway and elevators—would grind to a screeching halt. As a city of trade, thousands of tons of goods move in and out of its ports and terminals each day—by rail, truck, sea, and air. And as one of the world's most densely populated urban areas, it relies on communal delivery of services to an extent few cities do—

e.g., on a unified system of water delivery, on in-city generation of power, on the world's largest central steam system.

The magnitude and scope of the infrastructure that supports daily life in New York makes it the ideal subject for a study of how cities work. New York has everything: sewers, power, telecom, water, road, rail and marine traffic—all piled atop one another in what may be the densest agglomeration of infrastructure anywhere on earth. Exploring the systems that keep New York functioning at the pace it does provides a fascinating insight into the complexity of urban life at the dawn of the twenty-first century.

The chapters that follow explore five of the most interesting, and in many cases least visible, components of New York City's infrastructure: moving people, moving freight, providing power, supporting communications, and keeping the city clean. Like the essential systems that keep a human body running, each of these is vital to the functioning of the metropolis. And as with any lesson in anatomy, these complex systems—while interdependent— are best studied discretely. Each chapter is devoted to a system, and may be read as a whole or, alternatively, in sections designed to highlight its most important component parts. In either case, the graphic explanations and illustrations that accompany the text should form an integral—and we hope enlightening— part of the reading experience.

To Rebecca and Nathaniel

Hundreds of millions of people

live in the world's largest cities—places
like London, São Paulo, Shanghai,
and New York—and hundreds of millions
more commute to them each day. Wave
upon wave of vehicles course through
their streets, while thousands of tons
of cargo move in and out of their freight
terminals. Less visibly, millions
of gallons of clean water flow silently

MOVING PEOPLE

Between residents, visitors and commuters, tens of millions of journeys are made each day within New York City's boundaries. Many of these are made by mass transit, generally subway or bus; the remainder rely on taxis, private cars, or commercial vehicles. Just how the street and transit networks get everybody where they are going—safely, quickly, and with relatively little hassle—is one of the miracles of the modern city.

Streets are, of course, the most important element of moving large numbers of people—without a system of traffic signals and pedestrian crossings, urban life would be chaotic indeed. But subways are also important, and keep the volume of people on the roads to a manageable level. And bridges and tunnels, as an extension of the region's roadways, are equally necessary to move people smoothly across this city of islands.

New York is a city of streets. Almost 20,000 miles of streets and highways connect the inhabitants of the five boroughs. Only 1,250 of those miles represent highways: most are primary and secondary roads (7,300 miles) or local streets (11,000 miles).

The streets themselves, while simple in appearance, provide either the covering or the foundation for a world of related infrastructure. They protect the utilities and subway system below while providing a platform for traffic signals, parking signs and meters, streetlights, and sewers. Alongside the streets run the equally important sidewalks, which cater to pedestrian life and offer a foundation for conveniences like telephones and mailboxes, and for the urban vegetation known as street trees.

The street system we see today is both more complex and yet more orderly than at any time in New York's history. The earliest roads, clustered in lower Manhattan, were narrow affairs—easily choked with the traffic of the day. As Manhattan expanded northward, roads to northern settlements were developed somewhat randomly, probably along the routes of old Indian trails. Many if not all of these roads were the predecessors of today's broad north-south avenues.

The first real systemization of streets—and perhaps the event that best explains what we see around us today—was the development of the Grid Plan for Manhattan in 1811. Also known as the Commissioners' Plan, it fixed block and lot sizes and imposed the rectilinear grid that governs Manhattan's streets. While it succeeded in its primary purpose of underpinning orderly real estate development, it arguably failed to provide capacity for the heavy north-south traffic that would later be addressed by the development of the subways.

Streets

The Commissioners' Plan of 1811 The Commissioners' Plan for Manhattan, also known as the Grid Plan, was adopted in 1811 by the Common Council of the city. It mandated a gridiron layout for the expansion of Manhattan's street network, which up to that time had evolved haphazardly. East-west streets were spaced closely (the width of these streets was set at 60 feet between building lines). In contrast, north-south avenues were set farther apart and were wider (100 feet between building lines).

Keeping traffic moving on the streets of New York almost two centuries after the Grid Plan was introduced is a formidable and expanding task. From 1982 to 2000, when the population of the city increased by roughly 10 percent, the number of miles traveled within city borders rose by some 45 percent. Twenty years ago, roughly 3.4 hours each day were considered "rush hour"; today, rush hour has more than doubled, to between seven and eight hours each day.

But it is not just cars that are the focus of the city's traffic management efforts. Pedestrians are an equal presence on the streets, and any successful system must carefully integrate and synchronize the two. Doing so requires more than just the streetlights which govern traffic at 11,400 of the city's 40,000 intersections. It involves parking rules and restrictions, a pedestrian crossing system, one-way traffic patterns, and a host of other innovations—bus lanes, truck routes, "thru streets," and limited-access roads, to name just a few.

New York's Street Network

Highways
Major Roads
Minor Roads

Streets

Regional Traffic Most of the vehicles that clog Manhattan's streets, and many that regularly use streets in the outer boroughs, come from outside the city. Each day, an estimated 1.1 million cars and trucks cross into New York—from New Jersey, Long Island, or Westchester. Managing this flow—on highways, bridges, tunnels, and local roads—is an important part of the overall effort to keep New Yorkers on the move.

While the city's Department of Transportation (DOT) is largely responsible for the movement along city streets and bridges, a number of agencies share responsibility for the wider regional road transportation network, including the Port Authority (trans-Hudson bridges and tunnels); the Triborough Bridge and Tunnel Authority (Verrazano, Triborough, Whitestone, and Throgs Neck bridges, among others); and New York State (all state highways). Altogether, more than 16 different public transportation or safety agencies—with more than 100 different control rooms—operate within the region.

Transcom Screen Shot

Until recently, little if any coordination existed between these agencies, particularly with respect to infrastructure repairs, leaving the public to suffer through weekend after weekend of torturous road travel. That changed rather dramatically in the mid-1980s with the birth of Transcom, the "United Nations of transportation." An organization made up of 18 member agencies, Transcom monitors roadways in the tristate area and shares information on active and planned construction, sports events, and accidents with its member agencies. In addition to providing advice for broadcast over the radio, Transcom also orchestrates the variable message signs that drivers may find along the region's highways—telling them where the delays are and how they might best avoid them—and runs a round-the-clock control room to deal with major traffic accidents in the region.

Anatomy of an Accident When a major accident occurs in the metropolitan region, among the most important agencies to travelers is Transcom. Created originally as a way to enable its member agencies to share road construction information and resolve schedule conflicts, it soon gave birth to a round-the-clock control room, set up to communicate information about major traffic accidents in the region. The idea is simple: when an accident occurs, the transportation agency responsible for that road will be too busy dealing with the problem to inform neighboring jurisdictions of the incident. Transcom's job is to do just that.

Alert times

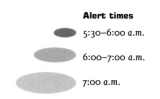

● 5:30–6:00 a.m.

● 6:00–7:00 a.m.

● 7:00 a.m.

At 7:00 a.m.
Subsequently, outreach was expanded with calls to agencies in Pennsylvania, Delaware, Maryland, and South Jersey. Communications referred to the fact that although the New Jersey-bound lower level was likely to have a lane clear shortly, delays would be extended due to the pending investigation, cleanup, and inspection for structural damage.

Agencies alerted: *New Jersey Transit Public Affairs, New York Waterway Ferry, Smart Routes (Boston), Metro Traffic (Providence).*

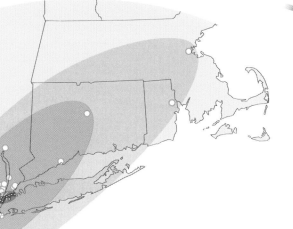

5:46 a.m.
*On Wednesday, June 6, 2001,
at 5:46 a.m., a truck fire closed the
New Jersey-bound upper level
of the George Washington Bridge.*

Before 6:00 a.m.
*Within ten minutes, the closure
was expanded to incorporate the
entire bridge.*

Agencies alerted: *Lincoln Tunnel,
Palisades Interstate Parkway, New
Jersey Transit Buses, Port Authority
Bus Terminal, Bergen County Police,
Fort Lee Police, NYPD Traffic*

*Management Center, MTA Command
Center, NYC DOT, NYC Transit
Buses, Shadow/Metro Traffic, Throgs
Neck Bridge, Whitestone Bridge,
Triboro Bridge, NY State Thruway,
LaGuardia Airport, New Jersey
Turnpike Authority.*

Before 7:00 a.m.
*The communications outreach,
following full closure, incorporated
the approaches coming into New
York. The information disseminated
gave the impression of a temporary
closure and estimated a delay of
60–90 minutes during the impending
rush hour.*

Agencies alerted: *New Jersey
State Police, Westchester County
Police, Port Authority Staten Island
Bridges, Verrazano Narrows Bridge,*

*Newburgh Beacon Bridge, Henry
Hudson Bridge, Tenafly Police, Leonia
Police, Edgewater Police, Englewood
Cliffs Police, Alpine Police, Palisades
Park Police, Pennsylvania DOT,
Connecticut DOT, New York State
DOT, JFK International Airport,
New Jersey Highway Authority,
Metro Traffic (Delmarva), Metro
Traffic (Hartford).*

Streets

Traffic Signals New York City is awash in traffic lights—11,400 of them to be exact. Contrary to the belief of many, they are not intended to limit speed; their primary purpose is to control right-of-way at intersections. In that sense, they are critical to the successful coexistence of people and cars in the urban environment. As with most conventional traffic lights, New York's have two phases: an east-west one and a north-south one. They generally operate on 60-, 90-, or 120-second intervals; the cycle is determined by local traffic conditions and may even be longer at times. The city's longest cycles are on the West Side Highway and on Queens Boulevard, each of which features cycle times of two minutes and fifteen seconds.

Anyone who has ever had a good run down Columbus or up Amsterdam avenues in Manhattan knows that lights on major avenues in the city are often set sequentially—turning green in about six-second progressions. (The ideal cruising speed is about 30 miles per hour, which is, not coincidentally, the speed limit.) But other lights turn green simultaneously, including on many of the two-way avenues like Park Avenue.

These lights are choreographed by New York City Department of Transportation at its Traffic Management Center (TMC) in Long Island City or manipulated manually by DOT staff. Connected to the lights by city-owned coaxial cable running under the streets, the TMC can change the length of red and green signals to accommodate daily fluctuations in traffic flow. During the morning rush hour, for instance, longer green lights on roadways leading into Manhattan facilitate the movement of inbound traffic; in the evening, the pattern is reversed. Similar adjustments are made for planned events, such as parades and ballgames, and for unplanned ones, such as roadway accidents or water main bursts.

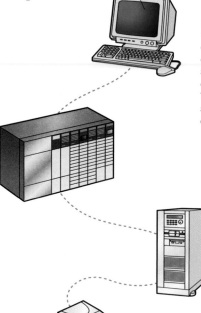

1. Fifteen computers at the TMC control up to 720 intersections each, monitoring real-time data including current signal displays, traffic detectors, and cycle lengths for each intersection. The area computers are connected to the intersections themselves by a variety of broadband cable technologies.

2. In addition to the computers, traffic flow is monitored on video cameras by TMC staff. Some 230 cameras, 90 of them in Manhattan, send images to the operations center. In case of an incident, the TMC engineers can remotely adjust signal timing online or, if needed, dispatch maintenance engineers.

3. The TMC also manages the "don't walk" signals that govern pedestrian flow. These are programmed to accommodate an average stride of four feet per second but can be adjusted to fit local conditions. In areas with a concentration of elderly people or young children, a slower stride (three feet per second) governs the pedestrian crossing cycle.

4. Although a green light on one face of a traffic signal generally corresponds with a red light on the opposite side, in most cases there is a two-second period when both sides are red.

5. Detector signals provide real-time information on traffic conditions. Magnetic loops adjacent to major intersections sense metal in cars passing above and send vehicle counts back to the operations center.

Traffic Light Buttons

Most of the 5,000 traffic lights not controlled directly by the TMC are set mechanically at boxes located near the intersection. Some of them—though not as many as one might expect—are still controlled, at least in part, by push buttons located on nearby poles. Called "semi-actuated signals" by traffic engineers, they first appeared in New York City in 1964. Located at the intersection of a major roadway and a minor side street, the idea was to allow traffic to flow freely on the larger road until a sensor in the side street—or a button located along it—signaled the presence of a vehicle or pedestrian.

Some 3,250 or so of these buttons remain in New York City, but fewer than a quarter of them actually work. The cost of removing the deactivated ones is high (roughly $400 per intersection), so they remain—a testament to a level of control by man over machine that many New Yorkers might wish still existed.

Split Phasing To address the daily conflicts between cars and pedestrians in New York, split phasing was introduced in parts of Manhattan in the fall of 2002 and became a permanent city initiative in 2004. Split phasing divides a traffic signal into three distinct parts, shown below, to provide pedestrians a safe street-crossing period, free from vehicle turns.

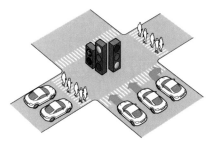

In the first instance, traffic on the avenue moves ahead on the green light, and traffic on the cross street is stopped.

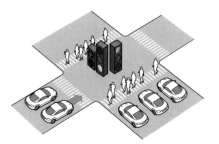

During the next phase, traffic on the avenue is stopped while vehicles on the cross street that are traveling straight ahead may proceed. Those turning from the cross street are not permitted to move, enabling pedestrians on the avenue to cross on both sides of the street.

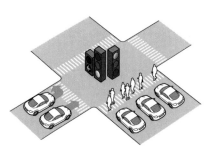

In the last stage, both cars turning from and going straight on the cross street are allowed to move; pedestrians may continue to cross on the nonturning crosswalk.

Streets

Traffic Cameras To monitor traffic flow across the city, DOT has installed cameras at major intersections and on highways and bridges. Many of these cameras simply allow traffic engineers to watch and adjust signal timing. Others are more active in their pursuit of drivers violating traffic rules. "Red-light cameras," for example, have been installed at fifty major intersections throughout the city. These cameras take high-resolution photographs of vehicles that go through red lights, including a close-up of the license plate. Summonses, including a photograph of the plate, are sent to violators.

New York was the first major city in the United States to implement a red-light enforcement program. Since its inception in 1993, more than 1.4 million summonses have been issued throughout the five boroughs. Only a small number have been contested, and very few ticket recipients have been found not guilty.

The program is apparently achieving its goals; studies have shown a 40 percent decrease in the total number of motorist violations at intersections with the cameras. City transportation planners would like to see the program expand to additional locations. In the meantime, another 200 locations have "dummy" cameras, which flash strobe lights in similar fashion to the real ones.

Traffic Camera Locations

Top Infraction Sites

1. Madison and East 79th St., Manhattan
2. Park Avenue and East 30th St., Manhattan
3. I-678 Service Rd. Eastbound at Hillside Ave., Queens
4. Rutland and Utica Aves., Brooklyn
5. I-495 Service Rd. Westbound at Van Dam St., Queens
6. Kings Highway and Remsen Ave., Brooklyn
7. 130th St. and 20th Ave., Queens
8. Avenue Z and Coney Island Ave., Brooklyn

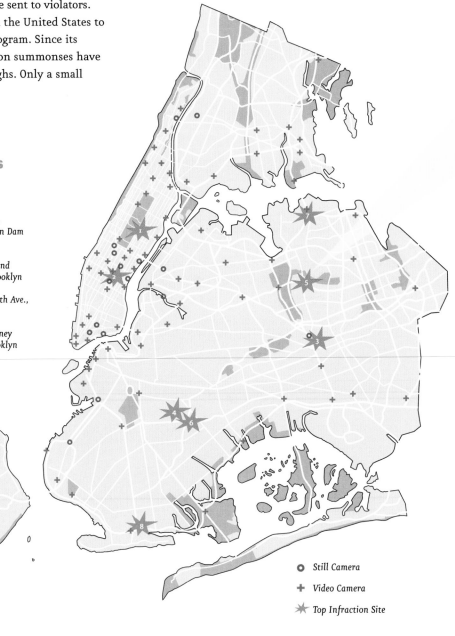

○ Still Camera
+ Video Camera
✳ Top Infraction Site

How Red-Light Cameras Work

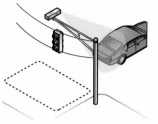

Red-light cameras are connected to the traffic signal and to two sensors buried in the pavement at the crosswalk or stop line. If a vehicle activates only one sensor after the light has turned red, the computer knows it has stopped at the edge of the intersection; if it activates both, the computer takes a digital photo of the car entering the intersection.

The computer calculates the speed of the vehicle and then takes a second shot of the car in the middle of the intersection. A camera records the date, time, speed, and seconds elapsed since the light turned red.

The license plate in the photograph is then referenced against Department of Motor Vehicles data, to ensure the plate matches the description on record. The data are then converted to a printed violation and forwarded to the city's Department of Finance, and a summons is sent by mail to the owner of the vehicle in question.

The digital or photographic evidence is stored online for a period of time in case the ticket is challenged.

Thru Streets Program

Midtown Manhattan auto speeds are notoriously slow—4.8 mph on average eastbound and 4.2 mph on average heading west. Many factors contribute to this problem beyond simply the high volume of vehicles using the streets: large numbers of pedestrians, illegal parking, construction activity, and truck loading are most to blame.

To better manage the midtown grid, in the fall of 2002, DOT selected certain streets to be designated "thru steets" to facilitate crosstown traffic. No turns would be permitted on or off five pairs of streets (36th/37th, 45th/46th, 49th/50th, 53rd/54th, 59th/60th) from Third to Sixth avenues —with the exception of Park Avenue—between the hours of 10 a.m. and 6 p.m. Neighboring streets were earmarked for localized circulation and commercial goods delivery, which was facilitated by providing curb space on both sides (as opposed to one side) of these adjacent streets.

Speeding Across Town

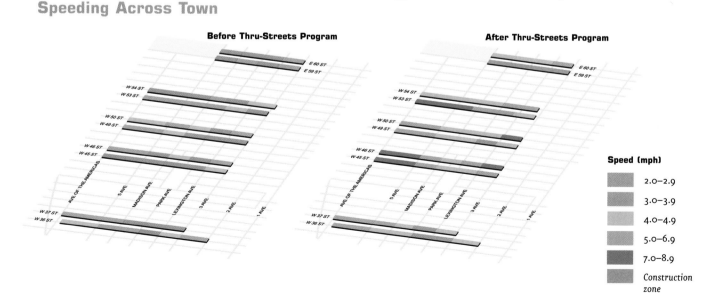

Before Thru-Streets Program

After Thru-Streets Program

Speed (mph)

- 2.0–2.9
- 3.0–3.9
- 4.0–4.9
- 5.0–6.9
- 7.0–8.9
- Construction zone

Streets

Traffic-Calming Measures

Thru streets and split phasing are just two of the newest weapons in the DOT's armory of traffic-management techniques; traffic lights and stop signs, in contrast, are two of the oldest. Some are labor-intensive, such as deploying police at congested intersections. Others, such as concrete barriers, may be temporary—and simply a way to protect or facilitate ongoing repair or construction.

Beyond these, there are more than a dozen other accepted "traffic-calming" measures designed to slow traffic or manage pedestrian flow.

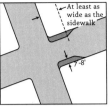

Neckdowns, *also called curb extensions, involve narrowing the street and widening the sidewalk.*

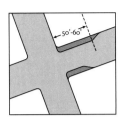

Bus Bulbs *involve widening the sidewalk at a bus stop so that buses do not leave the travel lane when stopping to pick up passengers.*

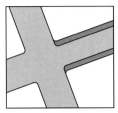

Roadway Narrowing *can be achieved either by widening the sidewalk or by using street markings to indicate narrowed travel lanes.*

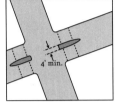

Pedestrian Refuges *involve small islands located in the middle of a two-way street, which allow pedestrians to cross in stages.*

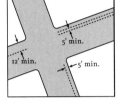

Bike Lanes *must be at least five feet wide when located next to a curb or parking.*

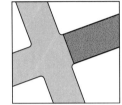

Roadway Color or Texture *may be used to accent or better define pedestrian crossing areas.*

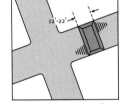

Speed Humps *may be as high as three or four inches, and may be circular, parabolic, or flat-topped in shape.*

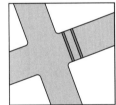

Raised Crosswalks, *two to four inches above the street, may be located at intersections or in the middle of a block.*

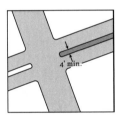

Roadway Medians *generally appear as raised islands along the center of a street.*

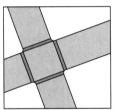

Raised Intersections *involve flat, raised areas that cover an intersection and often include a textured surface.*

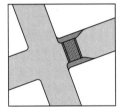

Gateway Treatment *involves a combination of measures, such as texture and raised street surfaces, to mark the entrance to a particular area.*

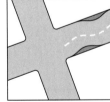

Chicanes *involve building out curb lines on alternating sides of the street.*

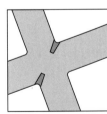

Partial Diverters *block travel in a particular direction at an intersection.*

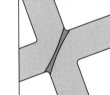
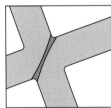

Diagonal Diverters *force all traffic to turn in a certain direction.*

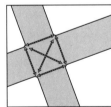

All-Pedestrian Phases *involve red lights on both streets at an intersection, which allows pedestrians protected crossing time.*

Leading Pedestrian Intervals *involve holding all vehicles at an intersection while giving pedestrians on at least one approach a green walk sign.*

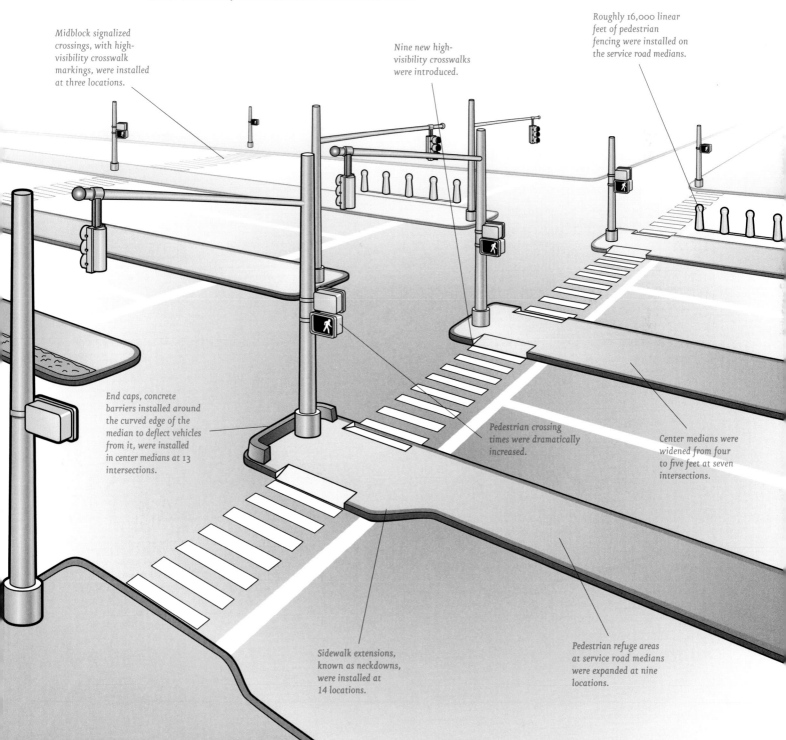

Calming Queens Boulevard
Twelve-lane Queens Boulevard is one of the widest streets in the city. It is also one of the busiest—and most dangerous. At least 50 pedestrians have been killed along Queens Boulevard in the last decade alone. To make the street safer, in 2000, DOT began implementing improvements along Queens Boulevard between the Long Island Expressway and Union Turnpike. Pedestrian fatalities subsequently dropped, from an average of nine a year in the 1990s to just two in 2005 and three in 2006.

Midblock signalized crossings, with high-visibility crosswalk markings, were installed at three locations.

Nine new high-visibility crosswalks were introduced.

Roughly 16,000 linear feet of pedestrian fencing were installed on the service road medians.

End caps, concrete barriers installed around the curved edge of the median to deflect vehicles from it, were installed in center medians at 13 intersections.

Pedestrian crossing times were dramatically increased.

Center medians were widened from four to five feet at seven intersections.

Sidewalk extensions, known as neckdowns, were installed at 14 locations.

Pedestrian refuge areas at service road medians were expanded at nine locations.

Streets

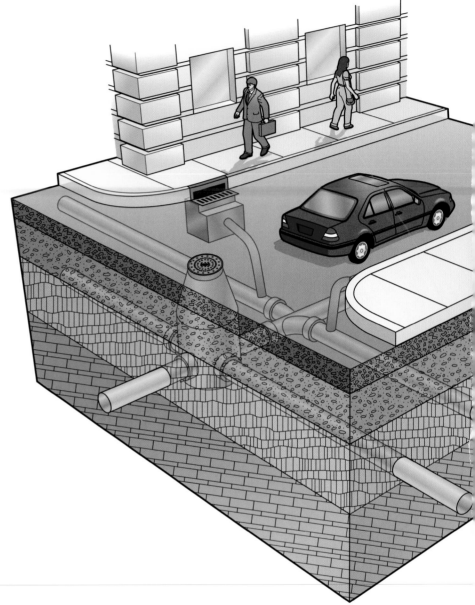

Street Surfaces For a city of extraordinary diversity, New York's streets are remarkably uniform—materialwise, that is. The vast majority of city streets consist of two layers of asphalt over a concrete base, although a handful are made entirely of concrete. Nearly all are graded to be slightly higher in the center, to allow water to run off into catch basins at street corners. Sidewalks are generally concrete with steel-faced curbs, though distinctive curbs made of granite or bluestone may be found within a historic district or near notable commercial buildings.

The use of asphalt as a paving material in the city is largely a twentieth-century phenomenon. Various forms of impacted stone and gravel were common street materials until 1872, when Battery Park and Fifth Avenue were the first streets to be paved with asphalt. But the longest-lived and most durable New York City street material to date is cobblestone: it has served as the road material of choice for two hundred years.

The Art of the Manhole Cover

The Catskill Water manhole cover was part of the second phase of the city's development of its upstate water delivery system.

Con Edison maintains the largest number of manhole covers in the city, encompassing various designs.

A commemorative manhole cover, called Global Energy, was designed by Karim Rashid for Con Edison in honor of the millennium.

The Borough of Richmond, now known as Staten Island, was created when the City of New York was incorporated in 1898.

Beneath City Streets

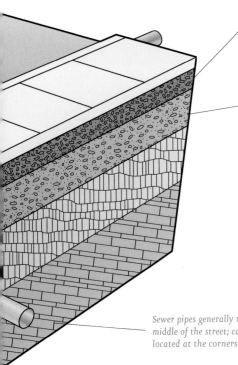

Most New York City roads have two top layers, each consisting of two to three inches of asphalt.

Below the asphalt is usually a base layer of concrete, though occasionally wood or clay is found below the surface.

Sewer pipes generally run down the middle of the street; catch basins located at the corners feed into it.

History in Stone

New York's cobblestones are not as old as most New Yorkers think. The concept of using small round stones as street paving dates back 350 years, but the cobblestones we ride across today are a mere 150 years old. Flat rectangles of Belgian granite, they were originally brought to New York in the 1830s as ship ballast.

Today, some 36 lane miles of cobblestones remain in New York City. Some streets, like Wooster, Greene, Mercer, and Bond in SoHo, are in a protected historic district; others—such as Perry and Bank streets—are not. Four times as expensive as asphalt, cobblestones in "unprotected" districts will only occasionally be replaced in kind. Often holes in these streets are filled in with asphalt or a mix of other kinds of stones.

This snowflake design dates from the late nineteenth century and was used to cover manholes used by the electric utility companies.

This cover can be traced back to the city's Department of Public Works.

This manhole cover design was done for the Fire Department.

The initials RTS stand for "rapid transit system" and are found above some subway shafts.

Streets

Street Repair New York City streets are in constant need of repair, either on a spot basis or requiring replacement of the street surfaces in its entirety. Most of the repair work is caused by extreme winter temperatures and heavy truck traffic, although some is simply a product of general wear and tear.

A variety of street defects keep DOT crews busy. Potholes are one kind of common defect; others include sinkholes, ditches, hummocks, ponding, open or failed street cuts, and cracked catch basins. Repair tactics vary, according to the defect. If the problem is the result of a failed utility cut, the responsible utility company is asked to make the fix. If the defect is too large for the DOT's emergency pothole crew, a temporary "make safe" repair is completed until the street can be properly restored. The most dangerous defects are those found in a crosswalk or driving lane —and they are repaired first.

Both street and utility crews rely on a uniform system of prerepair street markings to distinguish the type and location of underground infrastructure adjacent to a repair area. These seemingly random colored markings on streets are actually part of a sophisticated repair language: different colors and shapes are used to indicate either the limits of the work zone or the location of nearby utility lines. White paint is generally used to delineate a work site in advance of repair work being undertaken by city or private repair crews.

Street Defects

Ponding conditions refer to the buildup of water and occur at low points in the roadway as a result of poor drainage systems or insufficient grading.

Cave-ins, also known as sinkholes, are characterized by jagged edges around a deep hole.

Manhole covers can present a danger to drivers if they are cracked or missing or improperly placed above or below street level.

Old utility cuts are generally square or rectangular. If the repair was made within the last three years, the contractor is responsible for fixing it.

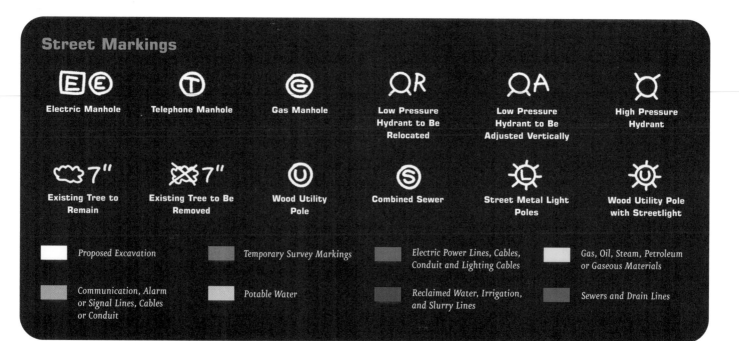

Street Markings

E E **Electric Manhole**	T **Telephone Manhole**	G **Gas Manhole**
R **Low Pressure Hydrant to Be Relocated**	A **Low Pressure Hydrant to Be Adjusted Vertically**	**High Pressure Hydrant**
7" **Existing Tree to Remain**	7" **Existing Tree to Be Removed**	U **Wood Utility Pole**
S **Combined Sewer**	L **Street Metal Light Poles**	**Wood Utility Pole with Streetlight**

Proposed Excavation

Temporary Survey Markings

Electric Power Lines, Cables, Conduit and Lighting Cables

Gas, Oil, Steam, Petroleum or Gaseous Materials

Communication, Alarm or Signal Lines, Cables or Conduit

Potable Water

Reclaimed Water, Irrigation, and Slurry Lines

Sewers and Drain Lines

Hummocks *are bumps in the roadway that result from heavy traffic and are often located near busy intersections.*

Potholes *generally display a bottom surface other than asphalt—usually dirt or gravel.*

Street hardware, *such as grates or vaults, may be misaligned with the road, cracked, or even missing.*

Open street cuts *are usually caused by power or telecom company work. They commonly exhibit colored markings on the asphalt.*

How Potholes Form

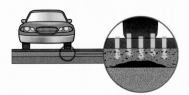

During cold weather, water in the pavement freezes and expands, breaking up the asphalt on and below the surface.

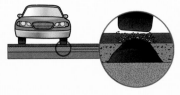

Melting ice leaves gaps, which are softened by water. The softened asphalt begins to come apart under the weight of passing vehicles.

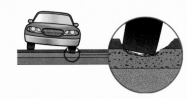

Broken pieces of pavement are displaced, leading to holes in the road.

How Smooth Are New York's Streets?

To measure the smoothness of city streets, test engineers drove a car across 670 miles of city streets. A "profilometer" on board measured the ups and downs associated with potholes, poorly aligned manhole covers, and inadequate repairs. Laser technology and citizen input were added to create the first set of reliable indicators of the "smoothness" of streets in the five boroughs.

Smoothness Score
Percent of blocks rated acceptable

- 80% or more
- 70–79%
- 60–69%
- 50–59%

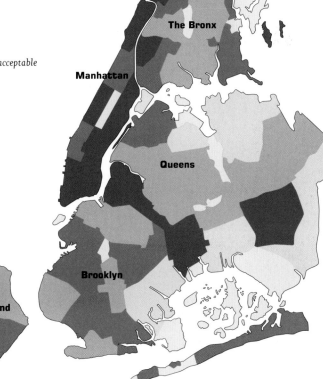

Streets

Rebuilding the FDR One of the most ambitious repair projects ever undertaken in New York City is under way on the east side of Manhattan—the rebuilding of the FDR Drive. Both upper and lower decks of the FDR will be rebuilt over a three-and-a-half-year period ending in 2007. To accommodate the 150,000 cars a day that use the highway, a temporary road over the river from East 54th to East 63rd St. was constructed. In addition to this dramatic roadway detour, a unique fendering system was designed and installed to protect the new road from collisions with passing ships.

E. 53 St.

1 During the first phase of work, a detour roadway is constructed 25 feet into the East River. This roadway is temporary in nature, so as to avoid any permanent restriction on channel shipping and to prevent any permanent impact on aquatic life.

2 Upon completion, northbound traffic is relocated from the lower level of the old roadway to the new roadway built over the river. Southbound traffic is shifted to the lower level of the existing highway.

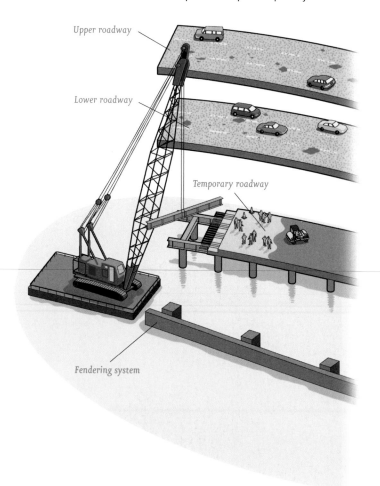

Upper roadway

Lower roadway

Temporary roadway

Fendering system

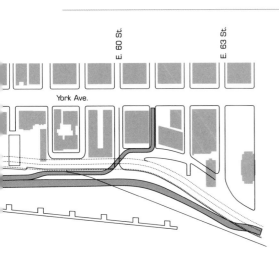

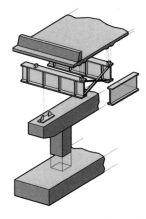

The temporary roadway detour, built over the river, is supported by a complex set of columns, bearings, stiffeners, and parapets.

3 In the third phase of work, southbound traffic is relocated to the upper level of the newly reinforced structure and work begins on the lower level. Northbound travel continues to use the new roadway over the river.

4 During the final phase of the rehabilitation project, the northbound traffic shifts back to the newly reconstructed lower level, and the temporary detour roadway and associated fendering system are removed.

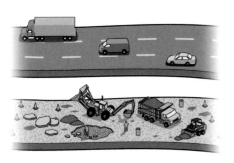

Streets

Sidewalks For many New Yorkers, sidewalks are a more important means of transportation than the streets themselves. Generally made of concrete but occasionally a more distinctive material, city sidewalks accommodate the millions of people—and also the street signs, parking meters, streetlights, trees, and trash cans—that give the streetscape its unmistakably urban flavor.

Every property in New York is required to have a sidewalk extending from the right-of-way line to the curb. It is the city's responsibility—one which it has largely lived up to—to install pedestrian ramps where pedestrian walkways intercept the curb.

Cellar doors *lead to belowground basements and open outward to prevent pedestrian mishaps.*

Sidewalks *are generally concrete, but may also consist of granite, brick, slate, marble, limestone, bluestone, or ceramic tile. "Distinctive sidewalks" must pass an aesthetics test, administered by the Art Commission, and an engineering test that considers their structural integrity and slip resistance.*

All sidewalks must **slope** *at a uniform grade from the property line down to the curb and longitudinally in the direction of the topography.*

Curbs *are required to be between five and seven inches deep. Generally made of steel-faced concrete, they may also be made of granite.*

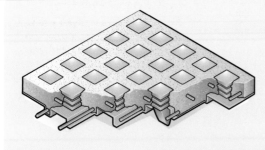

Sidewalk Vaults

Underneath many New York City sidewalks are vaults. At one time, most of these vaults were used for storage (primarily coal) by earlier generations of building owners. Although technically owned by the city and under DOT's jurisdiction, as well as that of the Department of Buildings, many have found more modern uses—living spaces, party rooms, offices, etc. Using vault space for other than storage requires **shoring up or replacing the vault lights that serve to illuminate the vault itself. Also known as sidewalk lights, these generally consist of glass prisms set into concrete with steel reinforcing bars. A prism shape is preferable to flat glass, as it diffuses and spreads light over a larger area; often multiple prisms set at different angles are used. All must meet the city's Department of Buildings load capacity ratings.**

Pedestrians Managing pedestrian flow is a top priority for city planners. Most intersections take into account pedestrian crossing patterns, and traffic signals are set accordingly. In general, pedestrian comfort is measured by how many people are on the sidewalk or waiting to cross at the corner. Where sidewalks are seriously over-crowded, a variety of measures are taken to mitigate the situation. These may include widening or lengthening the sidewalk, increasing the signal time, or—in extreme cases—building pedestrian bridges.

Grading Pedestrian Traffic

Volumes of pedestrian traffic vary street by street at a busy set of intersections, such as that in front of Lincoln Center in Manhattan.

Level of Service A: >130 sq. ft./ped.
At walkway LOS A, pedestrians basically move in desired paths without altering their movements in response to other pedestrians. Walking speeds are freely selected, and conflicts between pedestrians are unlikely.

Level of Service B: >40 sq. ft./ped.
At LOS B, sufficient area is provided to allow pedestrians to freely select walking speeds, to bypass other pedestrians, and to avoid crossing conflicts with others. At this level, pedestrians begin to be aware of other pedestrians and to respond to their presence in the selection of walking paths.

Level of Service F: <6 sq. ft./ped.
At LOS F, all walking speeds are severely restricted. There is frequent, unavoidable contact with other pedestrians. Cross- and reverse-flow movements are virtually impossible. Space is more characteristic of queued pedestrians than of moving pedestrian streams.

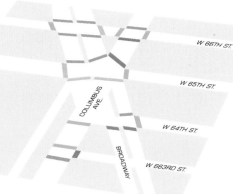

W 66TH ST.

W 65TH ST.

COLUMBUS AVE.

W 64TH ST.

BROADWAY

W 663RD ST.

Level of Service C: >24 sq. ft./ped.
At LOS C, sufficient space is available to select normal walking speeds and to bypass other pedestrians in primarily unidirectional streams. Where reverse-direction or crossing movements exist, minor conflicts will occur, and speeds and volume will be somewhat lower.

Level of Service E: >6 sq. ft./ped.
At LOS E, virtually all pedestrians would have their normal walking speed restricted, requiring frequent adjustment of gait. At the lower range of this LOS, forward movement is possible only by "shuffling." Insufficient space is provided for passing of slower pedestrians.

Level of Service D: >15 sq. ft./ped.
At LOS D, freedom to select individual walking speed and to bypass other pedestrians is restricted. Where crossing or reverse-flow movements exist, the probability of conflict is high, and its avoidance requires frequent changes in speed and position.

Streets

Street Signs Street signs abound in New York and are a key element in the city's efforts to manage its vehicular flow. Signs indicate where to turn and where not to, where to park and for how long, where to catch a bus, how fast to travel, etc. They also, of course, tell pedestrians or drivers what street they're on.

There are over one million signs on New York City streets, with parking and street-cleaning signs the most predominant. Just as a stop sign is universally recognized,

other street signs are also required to have a consistent shape, color, style, and meaning. Nearly all New York City signs are produced at DOT's sign shop in Maspeth, Queens.

Only very rarely does a New York City sign maker have a chance to be creative. The most recent opportunity came with the initiation of the "thru streets" initiative in midtown Manhattan in 2002. After much deliberation, the color purple—not yet a nationally designated color—was chosen for the new signs.

The Evolution of a Street Sign

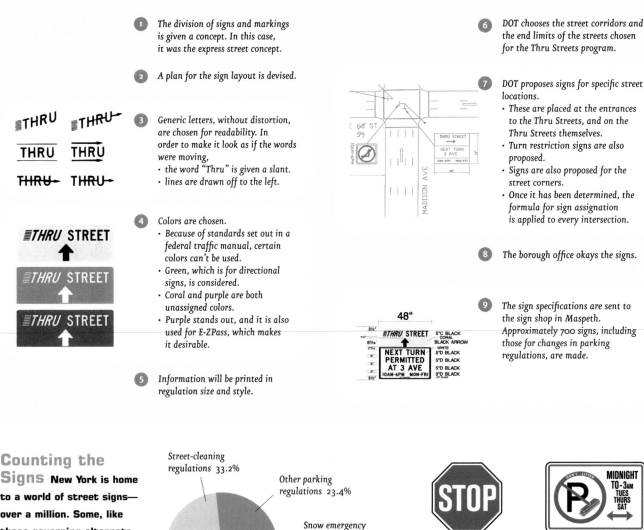

1. The division of signs and markings is given a concept. In this case, it was the express street concept.

2. A plan for the sign layout is devised.

3. Generic letters, without distortion, are chosen for readability. In order to make it look as if the words were moving,
   • the word "Thru" is given a slant.
   • lines are drawn off to the left.

4. Colors are chosen.
   • Because of standards set out in a federal traffic manual, certain colors can't be used.
   • Green, which is for directional signs, is considered.
   • Coral and purple are both unassigned colors.
   • Purple stands out, and it is also used for E-ZPass, which makes it desirable.

5. Information will be printed in regulation size and style.

6. DOT chooses the street corridors and the end limits of the streets chosen for the Thru Streets program.

7. DOT proposes signs for specific street locations.
   • These are placed at the entrances to the Thru Streets, and on the Thru Streets themselves.
   • Turn restriction signs are also proposed.
   • Signs are also proposed for the street corners.
   • Once it has been determined, the formula for sign assignation is applied to every intersection.

8. The borough office okays the signs.

9. The sign specifications are sent to the sign shop in Maspeth. Approximately 700 signs, including those for changes in parking regulations, are made.

Counting the Signs
New York is home to a world of street signs—over a million. Some, like those governing alternate side of the street parking, are fashioned specifically for city streets.

- Street-cleaning regulations 33.2%
- Other parking regulations 23.4%
- Snow emergency route signs 3.3%
- Arterial signs 1.5%
- Priority regulatory signs 13% (stop signs, do not enter)
- Turn restrictions and other intersection signs 6.3%
- Street name signs 19.3%

There are 130,000 "priority regulation" signs, which include stop signs as well as one-way arrow and "do not enter" signs.

An estimated 332,000 signs depicting street-cleaning regulations are found on city streets.

10   Street-marking changes, such as turn lanes and "thru" lanes, are made to the streets. Changes in parking regulations are also instituted.

11   The signals division installs mast arms for the signs.

12   Because all signs cannot be hung simultaneously, signs are hung in advance and then covered until the program is officially unveiled.

13   Inspectors from the planning unit are sent to study compliance and to make traffic counts. The Police Department also sends officers to monitor the traffic.

14   Small adjustments, such as allowing turns onto Park Avenue, are made. So as to not have to make entirely new signs, overlays are used when possible.

15   Maintenance records are begun. Signs last approximately 10 years.

Over 190,000 street signs can be found across the five boroughs.

There are 234,000 parking regulation signs.

More than 60,000 signs govern turns and other movements at city intersections.

Streets

Two Centuries of Streetlights

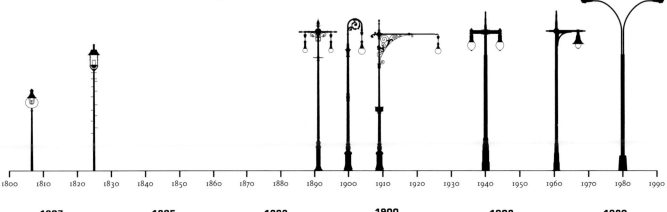

1807	**1825**	**1892**	**1900**	**1908**	**1980**
Among the earliest lampposts were those made of wood, which burned oil.	*Gas streetlighting was first installed in New York City in 1825 and continued into the twentieth century.*	*The first ornamental lamppost was placed on Fifth Ave. in 1892 and was limited to that street.*	*The bishop's crook, which appeared at the turn of the last century, was the second ornamental electric streetlight. It was reproduced in 1980.*	*The boulevard system was designed for streets with center malls. Mast arm lights reached over the roadway.*	*The original cobra streetlight featured a pendant light with the recognizable octagonal pole.*

Streetlights   A century after electric lighting began pushing gas streetlights into oblivion, New York City is awash in streetlights—333,670 in all. These include 35 to 40 different types.

The standard is the cobra streetlight, recognizable by an illumination component closely resembling the head of a cobra. First introduced in the 1950s, its design is purely functional; as a result, it has relatively few fans. More beloved are the 30 or so other models that survive in small numbers across the city—models with magical names like bishop's crook, reverse scroll, and lyre.

Historic or modern, street lighting is big business. The city pays Con Ed roughly $50 million each year in lighting bills, most of which finds its way to the New York Power Authority, which provides the electricity. In residential areas, illumination is at 110 volts; in commercial areas— where some business districts choose to buy and maintain their own streetlights— it is provided at 220 volts.

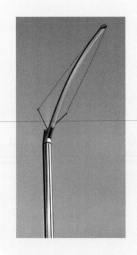

The Streetlight of the Future In February 2004, a competition to design a new citywide streetlight was undertaken by the city's Department of Design and Construction in conjunction with DOT. Some 201 entries from 24 countries were submitted to a panel made up of prominent architects, engineers, and public servants. The winning design, announced in October 2004, was submitted by Thomas Phifer and Partners. The new design will be used to light streets, sidewalks, and parks within the city.

Parking Meters

Parking meters act as a sort of traffic cop, regulating who can use valuable curb space and for how long. For the city, they are an important revenue generator: the 66,000 meters belonging to DOT's Division of Meter Collections collectively bring in upward of $70 million each year. Although parking meters in some areas have been replaced by "muni-meters," which issue tickets that must be placed on car dashboards, the familiar gray boxes are unlikely to disappear from the New York City landscape anytime soon.

Alternate Side of the Street Parking

Not all parking in the city costs money: in residential areas, parking is generally free to those lucky enough to find a space. But in many areas of the city even free parking is complicated, thanks to alternate side of the street regulations, which force drivers to clear one side of the street every couple of days for street cleaning. Begun in the 1950s on the Lower East Side as an experiment to facilitate the movement of newly mechanized street-cleaning machines, the program today incorporates some 10,000 miles of city roads.

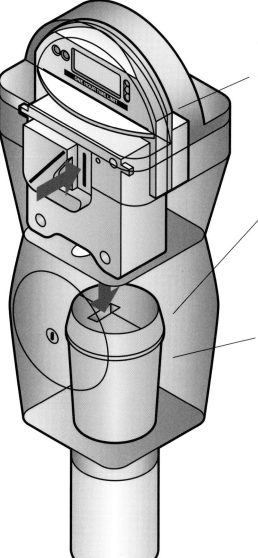

Most meters are designed to run a little long, to avoid challenges to their accuracy. A mechanical meter with a new timer can run from one to nine minutes long over the course of an hour.

Coins deposited in the meters accumulate in a coin box located in the "vault" of each meter. Separate keys are needed to open the vault and the coin box located within.

Each parking meter has the capacity to hold between $30 and $60, depending on the size of the coin box and the mix of coins.

DOT's Division of Meter Collections is responsible for collecting coins from each of the roughly 66,000 parking meters throughout the city at least once during a 24-day cycle. To collect coins from parking meters, collection crews use "canisters," steel boxes that roll along the streets. Field supervisors are assigned to observe the crews during their collection assignments to ensure collection procedures are being carefully followed.

Streets

 **Street Trees** Look down any street in the city, and chances are you will see at least a few trees sprouting from the concrete. At last count, there were an estimated 2.5 million trees in the city, of which 500,000 or so can be found on the streets (as opposed to the parks or backyards) of the metropolis.

It's not easy being green in the middle of the city. In addition to the usual challenges of disease and insects, street trees are also subject to vandalism, neglect, dogs, and generally difficult growing conditions. Yet they are a vital part of the streetscape, adding shade and color to the sidewalks, providing cleaner air, contributing to energy savings, and raising property values.

Street trees are one of the few municipal services where citizens may participate firsthand. New Yorkers can assist in tree maintenance so long as they complete the officially licensed, 12-hour "Citizen Pruner" course offered by Trees New York, which covers subjects such as tree biology and identification, pests, tree pruning, and tree-pit gardening.

Although New Yorkers can't own street trees (all trees planted in the city's right-of-way become city property after one year), they can plant them. There are several ways to do this:

- **Fill Out a Street Tree Request Form and Wait.** There is no charge for this service, but it can take up to two years until the requested tree gets planted by the Parks Department.
- **Visit the Parks Department's One-Stop Tree Shop.** Residents pay for the tree and its installation, but parks staff pick it, plant it, and care for it.
- **Plant It.** This requires a permit from the Borough Forestry Office, adherence to a list of approved species, and an inspection. If a new tree pit is to be dug, both a permit from the Department of Transportation and strict observance of guidelines for removing concrete are required.

Mapping New York's Trees

A Sampling of New York's Street Trees

Serviceberry is a small tree that produces white flowers.

Japanese Flowering Cherry is a small, rounded tree that does best in lawns and grassy strips.

Korean Mountain Ash is a narrow, small tree that produces a white flower.

Japanese Tree Lilac has a pyramidal shape and produces a white flower.

American Hornbeam is a slow-growing tree with a pyramidal shape.

Chinese Elm features purple leaves in the fall and is sensitive to the Asian long-horned beetle.

Red Maple is a medium-height tree with a rounded shape.

Bald Cypress has a pyramidal shape and can grow beyond 50 feet in height.

Gingko is a narrow, slow-growing tree that features yellow leaves in the fall.

Callery Pear grows to between 35 and 50 feet in height and produces a white flower.

English Oak is a slow-growing tree that can tolerate salty and dry conditions.

Shantung Maple is susceptible to the Asian long-horned beetle and is therefore prohibited from Queens, Brooklyn, and Manhattan.

European Ash is prohibited in Queens, Brooklyn, and Manhattan due to its sensitivity to the Asian long-horned beetle.

Golden Raintree is a rounded tree that produces yellow flowers.

Pin Oak has leaves that turn scarlet in the fall and can tolerate wet or dry soils.

Scholar Tree has a rounded shape and features cream-colored flowers.

of trains, it is absolutely the biggest—its 6,200 cars servicing 25 lines dwarf the fleets of even its largest competitors. And with 45,600 employees, represented by 25 unions, it is arguably one of the most complex subway systems to operate.

The system we recognize today by its award-winning color-coded map dates back to the nineteenth century. The earliest public transport within the city's bounds took the form of a 12-seat stagecoach running north along Broadway from the Battery

New York City's subway system is among the busiest urban transit systems in the world.

Every day, it handles over 4.5 million passengers—which equates to roughly 1.4 billion passengers each year. In terms of volume, it is among the world's largest— surpassed only by Tokyo, Moscow, Seoul and Mexico City. With respect to the number

starting in 1827; the earliest railways— elevated ones—made their debut in 1868. Not long after that, the first subway— an experimental one run on pneumatic power—was built furtively under City Hall, but was abandoned for lack of political support within just a few years.

The true predecessors of the modern subway were private subway lines set up by entrepreneurs after the turn of the last century. The earliest was the Interborough Rapid Transit (IRT) Line, which opened

Subway

Though granted a license to build a pneumatic tube to carry packages under Broadway between Warren and Cedar streets, Alfred Beach—a young inventor—proceeded to drill a "people-moving tunnel" in stealth.

In its first year, 400,000 people rode the car, but a stock market crash caused investors to withdraw and the service was terminated in 1873, three years after it began.

in 1904 and ran for nine miles along Broadway from City Hall to 145th St. in Manhattan. Initially calling at 28 stations, the IRT service was extended to the Bronx the following year, and subsequently to Brooklyn in 1908 and Queens in 1915. A second private line, the Brooklyn Rapid Transit Company (BRT), began providing service to Brooklyn at about the same time, but ran short of cash and emerged from bankruptcy as the Brooklyn-Manhattan Transit Corporation (BMT).

The first city-run service, the Independent Rapid Transit Railroad (IND), did not appear until 1932. Eight years later, with the private lines on the verge of bankruptcy, the city purchased both the IRT and BMT and became the sole operator of all subway and elevated lines within city limits. The city operated the lines under the jurisdiction of its Bureau of Transport until 1953, when the New York State legislature created the New York City Transit Authority as a separate public corporation to manage and operate all city-owned transportation—subways, buses, and trolleys.

Building a Subway Network

The earliest subway lines stretched along Broadway, from north to south, and connected the boroughs of Manhattan and Brooklyn. Gradually, the system expanded to Queens and the Bronx.

Introduction of subway lines by years

1900 1910 1920 1930 1940

Naming the Subways

The letter system of subway names was introduced following the unification of the independent subway lines in the **1940s**, and the system of color codes was introduced in **1979** to tie together, graphically, trains running along the same lines.

1, 2, 3, 4, 5, 6, 7, 9
The first subway company, the IRT, originally designated the subway routes by the avenues they ran under (e.g., Lexington) and included reference to the line's northern terminus. In 1948, this system was replaced with the numbered code that is still in use today.

A, B, C, D, E, F
The IND, which made its appearance in 1932, initially relied on a letter code that ran from A to H.

J, L, M, N, Q, R, W
The IND system of letter names was extended to the various lines of the BMT in 1960, using some of the rest of the alphabet.

AA, QB, RJ
For many years, express trains on the BMT and IND systems bore single letters while local trains on these lines carried double ones (the AA, for example, was the slow train to Harlem). In 1985, after color codes were introduced on the subway, double letters were eliminated entirely.

Subway

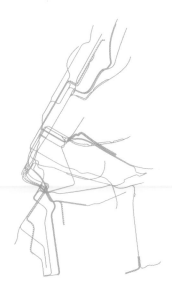

New York's subway runs across four boroughs day and night. It has more stations than other large systems and relies on a system of parallel express and local tracks to speed travel across the city.

Tokyo boasts an annual ridership approaching three billion—making it as widely used as the New York and Paris systems combined. It also reaches the highest speeds—62 miles per hour on certain stretches.

Moscow opened its first Metro line in 1935. Today, its 270-kilometer system carries roughly three billion passengers a year through 165 stations, some of which are deep enough to serve as bomb shelters in the event of a nuclear war.

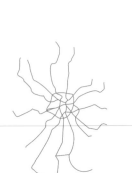

London's subway system is the oldest of the four major city subways in the world—it opened in 1863. It is also the longest, serving 253 route miles. Its longest line, the Central Line, stretches a full 46 miles west to east across London.

The Subway Network New York's subway system is made up of a far-flung and complex network of physical facilities. Its 25 interconnected lines stretch across four boroughs (Staten Island has its own overland railway system), with the longest (the C line) covering a length of over 32 miles. Some 68 bridges and 14 tunnels carry track through or into 468 subway stations. There, another 60 elevators and 161 escalators provide service to passengers.

The system runs along 842 miles of track—enough to stretch from the city to Chicago. Roughly 20 percent of it, or 180 miles, is not used for passenger service at all: it comprises support yards, shops, and storage areas that support the passenger network. But at its heart lie 660 miles of working track—two-thirds of it underground, with the remainder either elevated (156 miles) or at grade (57). Many of these miles lie in parallel, to support the local/express system—only 230 distinct "route miles" exist.

	Moscow	Tokyo	New York	London
Passengers per year	3.2 billion	2.7 billion	1.4 billion	886 million
Miles of routes	164.8	181	230	253
Number of stations	165	276	468	275
Number of cars	1,800	3,609	6,400	3,954
Hours of operation	6 a.m.–1 a.m.	5 a.m.–12:15 a.m.	24 hours	5 a.m.–1 a.m.

In many ways, the system—although largely underground —relies on technology similar to other American railroads. Its track gauge, the distance between the rails, is the same as all major American rail systems (4 feet 8 inches) and its signaling system is hardly unique. But what set New York's subway system apart from its very earliest days was the integration of local and express tracks into one network: New York was the first major world city to construct and operate such a two-tiered system. Today, its uniqueness stems largely from the 24/7 nature of its operation: few other major cities boast service all night long.

Morning Rush Hour

Scheduled minutes between trains during morning rush

9 minutes — B M

8 minutes — G C N

7 minutes — R W Q

6 minutes — D V

5 minutes — 2 3 J Z 5 A

4 minutes — 4 E F L

3 minutes — 1 9 6 7

2 minutes

Around-the-Clock Service: People per Car per Hour

Subway

Stations New York's subway system has more stations than any of its worldwide counterparts—468 in total. More than half of them are underground; the remainder are either elevated (153) or built on an embankment or "open cut" (39). The highest station is Smith/Ninth Street in Brooklyn (F, G lines); the lowest is 191st St. in Manhattan (1/9 lines), at 180 feet below street level.

Most New York subway stations are built to accommodate large volumes of passengers:

the network currently includes 734 token booths and over 31,000 turnstiles. The busiest single station in the system is Times Square: it handles over 35 million paying passengers each year. But that number is dwarfed by the number of passengers moving through the 34th St.–Penn Station –Herald Square complexes on the A, C, E, B, D, F, N, Q, R, and 1, 2, 3, 9 lines: collectively these three stations handle 60 million fare passengers each year.

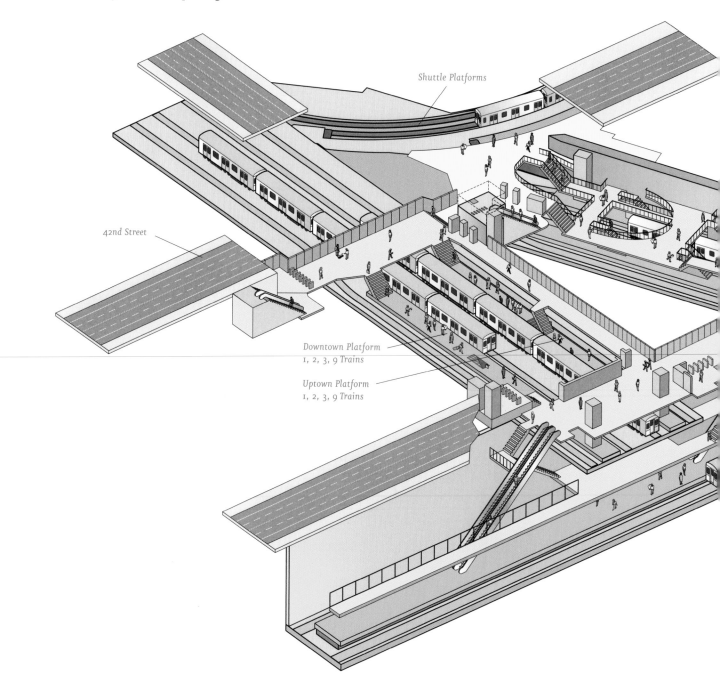

Shuttle Platforms

42nd Street

Downtown Platform
1, 2, 3, 9 Trains

Uptown Platform
1, 2, 3, 9 Trains

Times Square Station

The Times Square station, deep under West 42nd St. and Eighth Ave., is the busiest—and arguably the most complex—station in the subway system. The addition of new IRT, BMT, and IND lines to the original 1904 IRT station over time created the transportation hub that we know today. Major redevelopment currently under way at the station, costing $250 million, encompasses renovation to platforms, passageways, and mezzanines, as well as elevators and escalators.

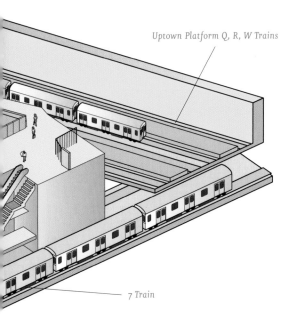

Uptown Platform Q, R, W Trains

7 Train

● Unused stations

○ Unused levels

✕ Unused platforms

Sedgewick Ave.

Jerome Ave.

91st St.

18th St.

Worth St.

City Hall

Cortlandt St.

Court St.

Myrtle Ave.

Abandoned Stations

Over the first hundred years of its existence, a number of subway platforms, levels, and even entire stations have been taken out of use. The most notable is the City Hall station on the original IRT line—known for its soaring ceiling, skylights, and period chandeliers. The sharp curves along the platform proved too much for subsequent generations of trains, and it was abandoned in 1945.

In total, there are nine abandoned stations in the network —five of which can today be seen from passing subway trains. These include West 91st St. along the 1, 2, 3, 9, lines; East 18th St. on the 4, 5, 6; Worth St. on the 4, 5, 6; the old City Hall station on the 6 line; and Myrtle Ave. on the B, D, N, and Q lines.

Subway

Token Timeline

The first token, minted by the IRT in **1928** in expectation of a two-cent increase, was never issued. The fare increase was overturned by the Supreme Court and the tokens went into storage until 1943, when they were sold to the Hudson and Manhattan Railroad for their metal value.

Upon unification of the IND, BMT, and IRT systems in **1940**, a transfer token was minted to enable passengers to move through a turnstile at no additional cost on the second leg of their journey by bus or subway.

The first full fare token debuted in **1953**, when the Transit Authority was created. At 16 millimeters in width, it featured the unusual "y" cutout. Fares stood at 15 cents through 1966 and then rose to 20 cents in 1970.

In **1970**, a new, larger token —still with the "y" cut— was introduced for the fare increase to 30 cents. This token survived the increase to 35 cents in 1973 and 50 cents in 1975 (despite an announcement, no doubt designed to avoid hoarding, that a new token would be introduced in 1975).

After 10 years, in **1980**, the 23mm "y" cut was replaced with a solid brass token (and a 60-cent fare). It remained in use for only six years, during which time the fare jumped to 75 cents, 90 cents, and, ultimately, $1.00.

1910   1920   1930   1940   1950   1960   1970   1980

Entering a Station

Trains usually enter the station at about 25 mph, under either a green or yellow signal. The train operator reverses the controller (motors run in reverse) to decelerate, and then employs air brakes to bring the train to a halt.

Before opening the train doors, the conductor (generally in the center car) lowers the window and points his or her finger when the car is lined up with a mark at the center of the platform. If the train is not lined up properly, he or she calls the operator to adjust the position of the train before opening the doors.

The train doors remain open for at least 10 seconds, while the conductor makes any announcement over the public address system. Before closing the doors, he or she must also announce "stand clear of the closing doors, please." The back half of the train doors are closed first, followed by the front.

The conductor then signals the train operator that he or she can proceed. The conductor then visually inspects both front and back sections of the train to ensure that no passengers are being dragged along by the train.

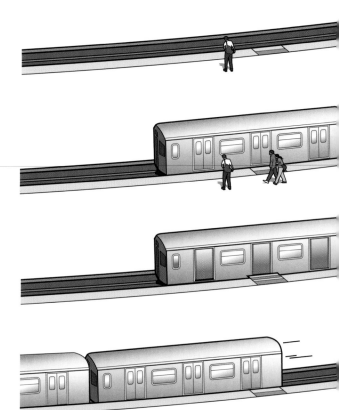

In **1997**, the MetroCard was introduced. In an attempt to win over customers, the MetroCard machines were designed to allow the customer to hold on to his or her credit card throughout the entire purchase (unlike with a bank ATM card). The buying process is intended to mimic a store dialogue, where people don't pay with a card until they have decided what to buy.

1990   2000   2010

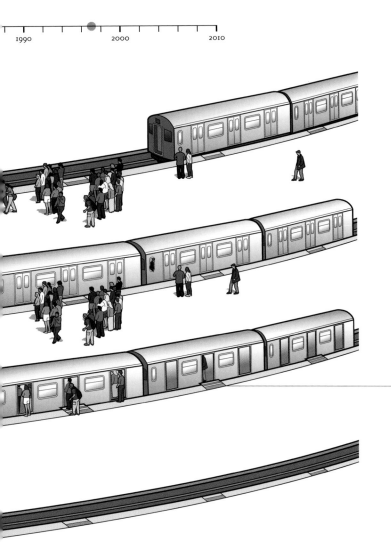

Subway Announcements

Although they are not always audible or intelligible, station announcements are an inevitable part of the subway experience. In general, announcements across the system are uniform: "Please do not hold train doors open"; "Due to a schedule adjustment, we are holding this train in the station"; or "There is a southbound train approaching 96th St." And with good reason: the Transit Authority's policy has been to provide only information needed to use the system wisely and "to cause minimal intrusion on our customers' right to think their own thoughts as they ride our trains."

Ad-libbing, as it is referred to in the "blue book" of subway announcements, is frowned upon. However, in an attempt to be more customer-friendly, announcers are permitted to use their discretion in making the following announcements "in order to add just a touch of something extra":

- The time of day: "Ladies and gentlemen, the time is three o'clock."
- Patronage recognition: "Thank you for riding with MTA New York City Transit."
- Or both: "Ladies and gentlemen, the time is three o'clock. Thank you for riding with MTA New York City Transit."

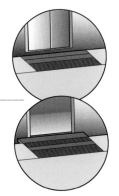

Gap fillers are used in stations with highly curved tracks, such as South Ferry, to bridge the space between the platform and the train car doors.

Subway

Trains Beginning in the late 1990s, the MTA placed a series of orders for new subway cars to replace what subway followers know as the "redbird" fleet—some of which had been in operation for 50 years. Built by Kawasaki and Bombardier respectively in Yonkers and upstate New York, the first trains were placed into passenger service in 2000. To date, more than 1,500 new cars have been ordered; deliveries continued through 2006.

The purchase of new trains is not as simple as it sounds. The maintenance shop at East 180th St. in Manhattan had to be overhauled to accommodate new maintenance routines. Extensive testing of the cars was undertaken on the Dyre Avenue line in the Bronx before the cars could move into passenger service. And delivery of the cars themselves was complex: those in Yonkers moved by flatbed truck to the 207th St. yard in Manhattan, where they were loaded onto the track, while the Bombardier cars arrived from upstate New York by train over the Hell Gate Bridge to the New York and Atlantic Railway's Fresh Pond Yard in Queens.

Dissecting the New Subway Car

The train's cab is now wholly computerized. A single lever governs traction and braking; alongside it on the control stand are a reversing key, a keypad, and an LCD flat-panel display. These are used to control doors and display train data.

While the new car is slightly smaller than the old due to thicker walls, new lighting and the removal of the backlit advertisements make it feel bigger. Shiny floors and the new color scheme also contribute to the feeling of a bigger space.

The new cars have slightly squarer ends and feature clear windows, which allow passengers to see into the next car.

Cars have been designed with nothing below the seats, for easy sweeping and maintenance. A dark pattern— black with speckles—was chosen for the floor, to hide dirt.

To improve access for wheelchairs, vertical poles between seats have been removed and a lift-up seat for wheelchair parking has been introduced. A ceiling-mounted bar was added to encourage taller people to hold on there, leaving room on the vertical poles for children and shorter people.

New LED signage appears both inside and outside of the cars. The outside signs indicate both train number and destination; on the inside, variable message signs and next-stop indicators provide useful information to passengers.

Car bodies are all stainless steel. Passenger entry doors are a full foot wider than they are in the preexisting fleet.

BROOKLYN BRIDGE

6 TO BROOKLYN BRIDGE

Bright-colored benches have replaced the traditional "scoop" seats. Front- and back-facing seats are now opposite benches that are parallel to the sides of the train.

Stainless steel cars, introduced in the 1980s to combat graffiti, scratched easily. Melanine, a formicalike material, was reintroduced to hide scratches. In addition, a protective layer of replaceable film has been placed on window glass to prevent permanent scratching.

Retired Subway Cars

Subway cars, once retired, are stripped of all asbestos and small metal items; handholds are removed and sold as memorabilia to collectors. The hollowed-out car then meets an unusual fate: it moves by barge south along the Atlantic Coast to be dumped on artificial reefs off the coast of Delaware, South Carolina, Virginia, or Georgia. Commercial fishermen are all too happy to get the cars: the reefs—piles of armored personnel carriers, tanks, and demolition debris—attract small sea mollusks and, in turn, increasingly large game fish.

Subway

Signals and Interlocking The subway relies on a century-old system of signals, known as "wayside color-light block signaling." Signals are located to the sides of the tracks and rely—much like traffic lights—on a system of red, yellow, and green signals to determine safe passage. Unlike the street system, which relies on traffic lights changing on a predetermined schedule, subway signals are determined by a system of track circuits and blocks, which detect the presence of trains on various portions of track. And while streetlights turn from green to yellow to red, subway signals go in reverse: from red to yellow, and from yellow to green.

There are essentially two types of signals that govern the operation of subway trains: automatic and approach signals. Automatic signals, the ones most frequently glimpsed by passengers riding on the system, are determined "automatically" by the presence or absence of a train on a length of track ahead. The distance along the track measured by the automatic signal is its "control length." Only when the control length is fully clear of a previous train will the signal turn to green, enabling the next train to move on to that portion of track.

Control lengths and signal placements almost always overlap beyond the next signal, and are designed so that an out-of-control train will stop before hitting something. Stopping relies on a system of automatic train stops, which trip a train violating a signal. The trip stop is a T-shaped metal rod, painted yellow, which goes up when a train runs a signal. It engages a trip cock on the wheel frame of the train, which in turn cuts power abruptly to the engine and applies the train's brakes in an emergency position. All cars, not just the lead car of a train, are equipped with trip stops.

Approach signals, in contrast, control the movement of trains across switches. They are generally characterized by two sets of vertical displays—one stacked above the other. These signals are not automatic: they are set at red by a terminal operator in a remote control tower until he or she determines that it is safe for the train to proceed through an area where tracks merge.

Proceed: the next signal is clear.

Proceed with caution: prepare to stop at next signal.

Stop: operate automatic release, then proceed with caution and be prepared to stop within vision.

Approach at the posted speed; continue on the main route. (A double signal generally has two sets of signal lights which control the movement of trains through switches and which are normally red until cleared by the tower operator.)

Gap filler is extended. Stop and stay.

Gap filler retracted. Proceed.

Wheel detector is on for the route. The switch is set to the diverging route and the train speed is within the speed limit required.

These areas where tracks meet, join, or switch—known as "interlockings"—are among the most complicated parts of the subway's signal system. Historically, the operation of interlocking was done by machines with mechanical levers that remotely controlled track switches and signals; these levers were designed to physically "interlock" with one another to avoid unsafe track configurations. Levers that related to a particular set of switches at a crossover could be thrown from the normal position (go straight) to the reverse position (switch track) by an operator. Other levers could force signals to red, to accommodate a crossing train, but an operator could not force them to green: the lever could instead be set to "permit the signal to clear," if the automatic signal did not detect the presence of a train on nearby track.

Interlocking technology remains at the heart of today's subway. Dozens of satellite towers house interlocking machines; in most cases, they are accompanied by a large model board displaying the track layout and featuring red lights to indicate the presence of trains on particular sections of track. To this day, tower operators have no automatic method of knowing which train is represented by the lights; train operators must push a "Train Identification Pushbutton" at the station located before an interlocking to request safe passage through the interlocking that lies ahead.

A Network of Control Towers Satellite offices, or "towers," are responsible for controlling the subway's local switches and signals. Inside the towers, transit personnel monitor train movement with the help of electronic maps linked directly to the subway's signal system.

▲ Master towers
✳ Interlocking towers
◉ Satellites

Subway

Power   The subway system is, perhaps not surprisingly, New York City's largest single electricity customer. Each year, it consumes some 1.8 billion kilowatt hours of power—enough to light the city of Buffalo for a year. Most power is provided by the New York Power Authority, which draws it from hydroelectric, nuclear, and fossil-fuel plants in New York State; a small amount is provided by the Long Island Power Authority for the part of the subway system that operates on the Rockaway Peninsula.

Like the signal system, power delivery to the subway has changed little in concept since the system opened in 1904. Alternating current is sent from generators along high-tension cable to 214 substations along the various routes. There it is changed from alternating current to direct current (625v) and fed—via 900 miles of heavy traction power cables and 1700 circuit breakers—onto the third rail. Every train has a "shoe" connected to the third rail, which picks up the electricity and allows the train to move.

A separate system of power, involving an additional 1,600 miles of cable, provides alternating current to signals, ventilation and line equipment, and station and tunnel lighting. By separating the two systems, the lights remain on when power to the third rail is cut off and vice versa.

The amount of power provided to the system is largely a function of the distance covered by a particular line. Trains draw power according to their operational needs. Express trains generally travel at speeds averaging 25 miles per hour. Local trains operate at approximately 15 miles per hour on average below 96th St., and at 18 miles per hour above it due to the greater distance between stations.

Train Crews   In general, subway trains are crewed by two people: a motorman, or operator, and a conductor. The operator rides in the cab at the front of the train and governs the movement of the train along the tracks and through stations. He or she is entirely responsible for the safe operation of the train when it is moving.

The conductor, in contrast, rides in the middle of the train. He or she is responsible for the opening and closing of the doors of the train at stations and announcements. The conductor indicates to the operator when the train is appropriately aligned upon arrival at a station and is in charge of announcements relating to the boarding and exiting of passengers.

Making It as a Motorman

Only recently has anyone other than a subway employee been able to apply to drive an NYC subway train. At a test given in November 2003, some 14,000 people showed up at 14 locations across the city to compete for about 300 train operator jobs. The test itself had 70 questions and took about three and a half hours to complete. Sample questions included:

1. Safety rules are most useful because they:
a. Make it unnecessary to think
b. Prevent carelessness
c. Are a guide to avoiding common dangers
d. Make the worker responsible for any accident

Daily Power Usage

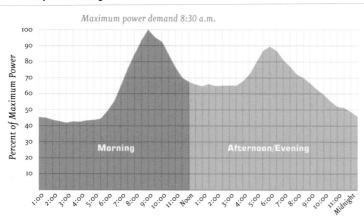

Maximum power demand 8:30 a.m.

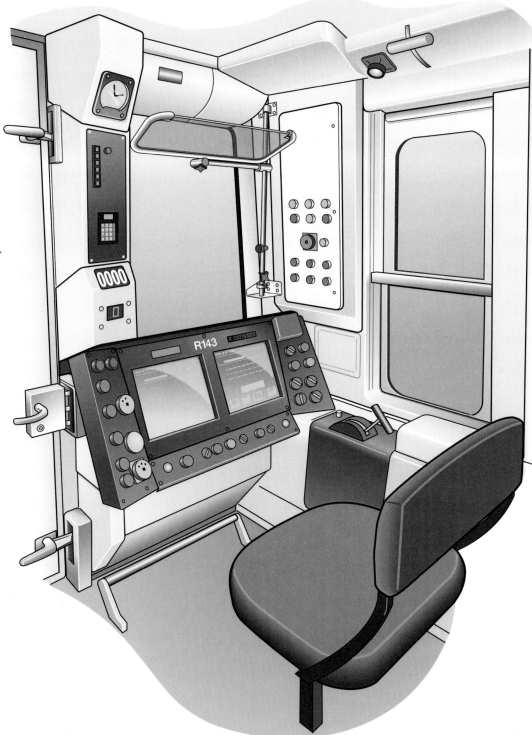

Inside the cab *The train status panel, located in front of the operator in the cab, provides information about all aspects of the train's operation, including location, all aspects of maintenance, the status of communication and signage, power and braking systems, and trouble indicators. The control panel includes a lever that governs the train's movement.*

2. The maximum speed permitted when a train is passing through a passenger station without stopping is:
a. 5 mph
b. 10 mph
c. 15 mph
d. Series speed

3. Third rail power is used to operate the:
a. Compressors
b. Emergency car lights
c. Motorman's indication
d. Conductor's signal
   lights

Answers: 1. c, 2. c, 3. a.

Subway

Breakdowns For a century-old system, New York's subways are pretty reliable. The "mean distance between failures"—the distance a car travels on average between breakdowns—averages over 100,000 miles. Nevertheless, there is hardly a subway rider who has not experienced at least one breakdown in his or her mass transit career. When such an incident occurs, a standard set of procedures is followed:

1. The Subways Control Center is notified, generally by a portable radio carried by train staff. The notification would include the train's "call signs" (e.g., "1427 C 168th St./Euclid" is the call sign for a train that departed 168th St. at 2:27 p.m. and is destined to terminate at Euclid Ave.), its location, and a description of the problem (often coded).

2. Subways Control Center notifies the local satellite tower and the tower begins rerouting service around the delayed train. Rerouting directions are a function both of the layout of the tracks in the area around the incident and of the location of switches enabling diversion from one set of tracks to another.

3. For many problems, particularly in heavily trafficked corridors, there are predetermined detour routes. For example, a problem on the A or C line in lower Manhattan would result in all A and C trains being diverted to the F line between Jay St. in Brooklyn and West 4th St. in Manhattan.

The Emergency Brake

There are a number of situations that can trigger deployment of a train's emergency brake. In any of these cases, a vent is opened, leading to a reduction in brake pipe pressure (normally at 110 psi); this triggers the immediate application of electropneumatic friction brakes.

The train operator can manually turn the brake valve to the emergency position.

Emergency Codes At least 11 emergency conditions are readily identified by subway radio code:

12-1	Emergency—clear the air
12-2	Fire or smoke on train or roadbed
12-3	Flood or serious water condition
12-5	Stalled train
12-6	Derailment
12-7	Request for assistance
12-8	Armed passenger
12-9	Passenger under train
12-10	Unauthorized person on track
12-11	Serious vandalism
12-12	Disorderly passengers

The passengers or conductor can pull the emergency brake cord located inside each car.

The train operator could become incapacitated and fail to keep the master controller lever depressed (the "dead man's" feature).

The trip cocks on an individual car could strike an object on the tracks.

Rerouting a Train New York is fortunate to have a subway system generally comprised of three or four tracks. Should there be a problem on the C line, for example ("Control 1427 Charlie 168 to Euclid is at 81st St. experiencing a door malfunction"), the nearest switches—in this case at West 125th St. and West 59th St.—would be identified. The local tower, at 59th St., would reroute all service from the local track to the express track at 125th St. At 59th St., local trains would switch back to the local track.

Subway

Rail Yards Keeping subway cars moving around the clock requires a far-flung network of support services, including subway storage yards and maintenance shops. Thousands of Transit Authority employees in the Car Maintenance Department are responsible for regular inspection of the cars at the 13 maintenance shops in Manhattan, Brooklyn, Queens, and the Bronx. Cleaning and repair also occur at these shops.

In addition, two overhaul shops are used for major repairs and car rebuilding: the Coney Island shop in Brooklyn and the 207th St. shop in Manhattan. The overhaul shop at Coney Island, for example, operates 24 hours a day to provide repair services to both the Transit Authority and to the Staten Island Rapid Transit Fleet.

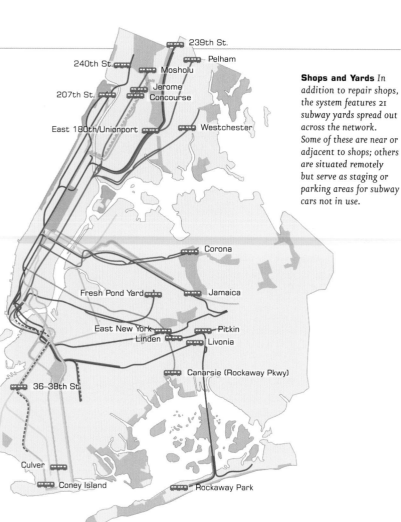

Shops and Yards *In addition to repair shops, the system features 21 subway yards spread out across the network. Some of these are near or adjacent to shops; others are situated remotely but serve as staging or parking areas for subway cars not in use.*

Subway Pumps Keeping subway tracks dry is key to the smooth running of the system. Each day, 309 pump plants—with a total of 748 pumps—remove up to 13 million gallons of rain and other water from the subway system. Fed by a system of drains running under the tracks, they pump water into a manhole under the street that empties into the city's storm-water system.

Flooding can, however, occur for a variety of reasons. Sometimes debris blocks the drain along the tracks. Other times the pipes leading from the drains to the pump rooms are overwhelmed with the volume of water that builds up on the track. To minimize failures, the TA is in the process of installing wider inlet pipes and placing slatted boxes over the track drains to better protect them from blockage.

Support Cars Passenger cars make up the bulk of the Transit Authority's fleet— close to 6,000 in all. But there are another 350 or so rail cars that are largely hidden from public view, without which the entire system would come to a screeching halt. Some of them—for example the refuse and revenue cars—move through stations to provide basic collection services. Others, like snow blowers and tank cars, are rarely seen inside the tunneled portion of the network. And still others—weld cars, crane cars, and signal supply cars, among others— are part of the fleet responsible for repair and maintenance across the system.

Revenue Collection Cars *There are 10 two-car trains that move through the stations collecting fares from the station booths. These trains will soon be phased out and replaced by armored trucks.*

Signal Supply Cars *These cars are equipped with cranes to remove existing signals from the track and install new signals, and are generally pulled by passenger cars.*

Weld Cars *These are retired passenger cars that are designed to carry eight 390-foot rails and transport them to areas where they will be installed on the track.*

Snow Blowers *These cars are equipped with a jet engine to blow snow off the tracks. They require two locomotive escorts as the unit is too short for the signal system to recognize its presence on the track.*

Crane Cars *These cars are generally used to carry, lift, or unload sections of track being replaced but also may be used to lift equipment such as generators and track ties. They are generally pulled by locomotives.*

Tank Cars *These cars are used to carry liquids around the subway system. They are generally pulled by locomotives.*

Locomotives *Locomotives pull nonpowered work cars, such as flat cars and crane cars, to and from job sites. Sixty-two are diesel electric locomotives, and ten are electric.*

Ballast Regulators *These cars spread ballast on the tracks after it is dumped. They feature a rotating broom which distributes ballast and sweeps excess ballast onto a conveyor for removal.*

Flat Cars *These cars generally haul machinery and other equipment to and from work sites. They can carry loads up to 30 tons.*

The Vacuum Car

Among the most unique of the specialized subway cars is the vacuum car. Weighing several tons and costing roughly $15 million, it sucks in 70,000 cubic feet of air per minute in a never-ending battle to keep the subway tracks clean. Carefully designed to steer clear of small, heavy objects (like track ballast), it is estimated to have picked up more than five million pounds of debris in the last two years alone. But even a smart machine like this one needs help at times: an advance team is sent out along the track ahead of the vacuum to pick up larger items, such as shoes, cell phones, and cosmetic bags, that might not be so easily digested by it.

New York's fabulous harbor and multiple waterways once made it a center of trade, but

today they make it a city of bridges and tunnels. Over 2,000 of them provide uninterrupted vehicular movement throughout the region. Seven agencies claim jurisdiction over this web of crossings:

Bridges & Tunnels

the Port Authority of New York and New Jersey, the Metropolitan Transportation Authority (MTA), DOT, New York State Department of Transportation, New York City Department of Environmental Protection (DEP), Amtrak, and the New York City Department of Parks.

Nearly all of the city's major bridges—and several of its tunnels—have broken or set records. The Holland Tunnel was the world's first vehicular tunnel, when it opened in 1927. The George Washington and Verrazano Narrows bridges were the world's longest suspension bridges when they opened in 1931 and 1964 respectively; likewise, the Bayonne Bridge, a steel arch structure connecting Staten Island with Bayonne, New Jersey, was very briefly the longest of its type.

New York's crossings date back to 1693, when its first bridge—known as the King's Bridge—was constructed over Spuyten Duyvil Creek between Manhattan and the Bronx. Composed of stone abutments and a timber deck, it was demolished in 1917. The oldest crossing still standing is Highbridge, which connects Manhattan to the Bronx over the Harlem River. Never designed to carry vehicles, it was opened in 1843 to carry water to the city as part of the new Croton Aqueduct system.

Ten bridges and one tunnel serving New York City have been awarded some degree of landmark status. The Holland Tunnel, operated by the Port Authority of New York and New Jersey, was designated a National Historic Landmark in 1993 in recognition of its pioneering role in vehicular tunnel technology. The George Washington Bridge (another Port Authority facility), Highbridge (operated by DEP), and the Hell Gate Bridge (operated by Amtrak) have also been made landmarks. So too have seven other

In 1927, after seven years of construction, the Holland Tunnel—the first mechanically ventilated vehicular underwater tunnel—opened. The toll was 50 cents.

More than 14,400 miles of steel cable were used in the construction of the Brooklyn Bridge, which opened to great fanfare in 1883.

bridges under the control of New York City DOT: the Queensboro, Brooklyn, Manhattan, Macombs Dam, Carroll Street, University Heights, and Washington bridges.

Today, the 14 major bridge and tunnel crossings that connect the city to its neighbors account for millions of vehicle passages each day. The busiest—the George Washington Bridge, the Verrazano, and the Triborough—were all designed by one man: an engineer named Othmar Amman. (Amman also designed the Whitestone and Throgs Neck bridges and the Lincoln Tunnel—making him something of a father figure in New York City civil engineering circles.)

E-ZPass E-ZPass is an electronic toll collection system now operating in eleven states in the Northeast and Midwest. Participating vehicles are equipped with a small electronic tag that transmits data to a remote computer at a customer service center in Staten Island. The data are processed and a toll is automatically deducted from the driver's prepaid account.

E-ZPass has been wildly successful since its introduction in 1993 and is now in operation at all MTA and Port Authority bridges and tunnels in the region.

Manhattan's Crossings

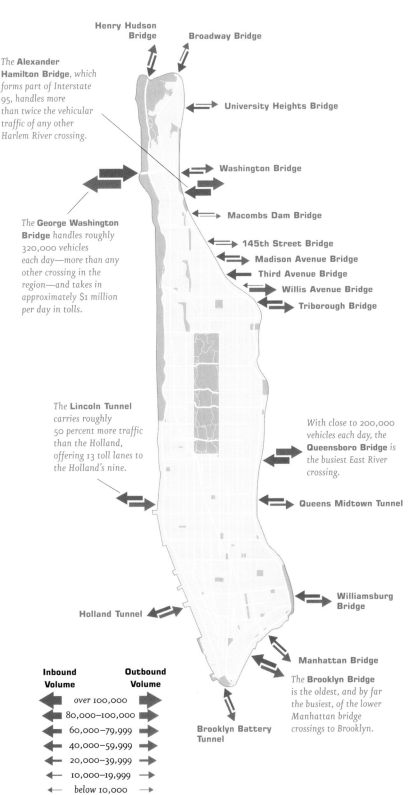

The **Alexander Hamilton Bridge**, which forms part of Interstate 95, handles more than twice the vehicular traffic of any other Harlem River crossing.

The **George Washington Bridge** handles roughly 320,000 vehicles each day—more than any other crossing in the region—and takes in approximately $1 million per day in tolls.

The **Lincoln Tunnel** carries roughly 50 percent more traffic than the Holland, offering 13 toll lanes to the Holland's nine.

With close to 200,000 vehicles each day, the **Queensboro Bridge** is the busiest East River crossing.

The **Brooklyn Bridge** is the oldest, and by far the busiest, of the lower Manhattan bridge crossings to Brooklyn.

Henry Hudson Bridge
Broadway Bridge
University Heights Bridge
Washington Bridge
Macombs Dam Bridge
145th Street Bridge
Madison Avenue Bridge
Third Avenue Bridge
Willis Avenue Bridge
Triborough Bridge
Queens Midtown Tunnel
Williamsburg Bridge
Manhattan Bridge
Holland Tunnel
Brooklyn Battery Tunnel

Inbound Volume		Outbound Volume
	over 100,000	
	80,000–100,000	
	60,000–79,999	
	40,000–59,999	
	20,000–39,999	
	10,000–19,999	
	below 10,000	

Bridges & Tunnels

Bridge Types

Cantilever bridges *exhibit a lacy superstructure of rods, plates, girders, and cross braces to support the bridge deck. The Queensboro Bridge is a notable example of a cantilever truss bridge in New York.*

Suspension bridges *depend on large cables that are strung over a pair of towers and anchored to shoreside blocks of concrete. Suspender cables are hung from the primary cables to hold the roadway. The George Washington, Verrazano, Brooklyn, Manhattan, and Williamsburg bridges are all notable New York City suspension bridges.*

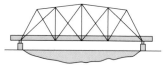

Trestle bridges *are supported by a series of connected pilings or beams.*

Girder span bridges *are used to bridge relatively short distances. The steel girders carry the roadway load to supports at each end of the bridge.*

Steel arch bridges *are made up of one or more arches made out of concrete or steel. Only one steel arch bridge is under city control: the Washington Bridge over the Harlem River, which is constructed from twin steel arches.*

Truss bridges *are characterized by road decks supported by steel trusses that rest on piers and abutments.*

Bridges New York features a variety of bridges—of all lengths and types, carrying everything from cars, trucks, and subway trains to bicycles and pedestrians. Technically, a bridge is defined as a structure that spans a distance greater than 20 feet. According to this definition, the longest municipally owned bridge is the Gowanus; the longest railroad-owned bridge is the elevated tracks in the Bronx by Yankee Stadium.

But the bridges that New York City is famous for are not those over land, but rather those that connect its islands. The Verrazano Narrows Bridge, the George Washington Bridge, and the Brooklyn Bridge are considered among the most beautiful in the world. Others fall into the category of the most functional. Consider the Williamsburg Bridge, for example: its two heavy rail transit tracks carry tens of thousands of people each day on the J, M, and Z subway lines; its eight traffic lanes support over 140,000 vehicles a day, and its sidewalk provides service to about 500 pedestrians each day.

In total, roughly 2,000 bridges dot the landscape of the city, over 700 of them under the responsibility of New York City's DOT. Twenty of these connect boroughs; the remainder are somewhat evenly distributed across them. Within this total are 25 movable bridges whose openings are governed by Coast Guard regulations and organized by means of two-way radios or a telephone call to the DOT's Division of Bridges.

Movable Bridges

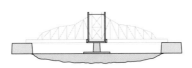

Retractile bridges *are mounted on tracks alongside a navigable waterway and are "retracted" when a ship needs passage. They were popular in the nineteenth century for narrow crossings requiring maximum horizontal clearance. The Barden Avenue and Carroll Street bridges are examples of retractile bridges.*

Swing bridges *are supported by a central pier situated in the water. The bridge is opened by rotating it horizontally along wheels on a circular track, and in its open position it forms two separate channels for passing vessels. The Third Avenue, Madison Avenue, and Macombs Dam bridges are all examples of this type.*

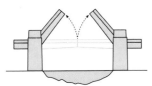

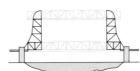

Bascule bridges, *also known as drawbridges, rely on counterweights to vertically lift two spans. The Pelham, Hamilton Avenue, and Greenpoint Avenue bridges are all bascule bridges.*

Vertical lift bridges *are movable bridges with roadways that can be raised in an elevatorlike fashion. This is accomplished through a system of attaching supporting end cables to rotary drums in towers on the sides of the waterway. The 103rd Street Bridge, Ward's Island Foot Bridge, and Roosevelt Island Bridge to Queens are examples of vertical lift bridges.*

Most Frequently Opened Bridges (1988–2002)

1. *Shore Road–Pelham Parkway (Bronx)*
2. *Hamilton Avenue (Brooklyn)*
3. *Ninth Street (Brooklyn)*
4. *Greenpoint Avenue (Brooklyn/Queens)*
5. *Metropolitan Avenue (Brooklyn)*
6. *Bruckner Expressway (Bronx)*
7. *Pulaski (Brooklyn/Queens)*
8. *Third Street (Brooklyn)*
9. *Mill Basin (Brooklyn)*
10. *Carroll Street (Brooklyn)*

- *land bridge*
- *waterway bridge*
- *railroad bridge*
- *movable bridge*

New York City's Bridges

Bridges & Tunnels

Brooklyn Bridge *The Brooklyn Bridge is the oldest suspension bridge in the harbor. When built, its towers were the tallest structures in lower Manhattan. It opened to great fanfare in 1883, having cost just over $15 million and 20 lives.*

Williamsburg Bridge *In contrast to the Brooklyn Bridge, which took 13 years to build, the Williamsburg Bridge was built in seven—opening to traffic in 1903. Its main span (1,600 feet) and height (135 feet above mean high water) are almost precisely those of its sister to the south.*

Queensboro Bridge *The Queensboro Bridge, the only major cantilever bridge in the region, was opened in 1909. It was originally configured to accommodate two elevated railway lines and two trolley lines— both of which were eventually removed—as well as a marketplace on the Manhattan side tiled in Gustavino tile, the same tile found in Grand Central Terminal's famed Oyster Bar.*

Bayonne Bridge *The Bayonne Bridge is one of the longest steel arch bridges in the world, with a midspan clearance of 151 feet. The hyperbolic curve arch over the roadway is complemented by steel trusses placed in a triangular pattern.*

George Washington Bridge
The George Washington Bridge opened in 1931 with one deck, though it was designed to handle a second level to carry either rail traffic or additional road traffic. In 1962, the second level was eventually completed—for vehicular traffic only. Today, its 14 lanes make it one of the world's busiest suspension bridges.

Verrazano Narrows Bridge
When it was opened in 1964, the Verrazano Narrows Bridge was the longest suspension bridge in the world. At 228 feet above the water, it also sat higher than any other bridge in the harbor; indeed, its towers were so high and so widely spaced that its builders had to account for the curvature of the earth's surface, which is why the towers' tops are 1 5/8" farther apart than their bases.

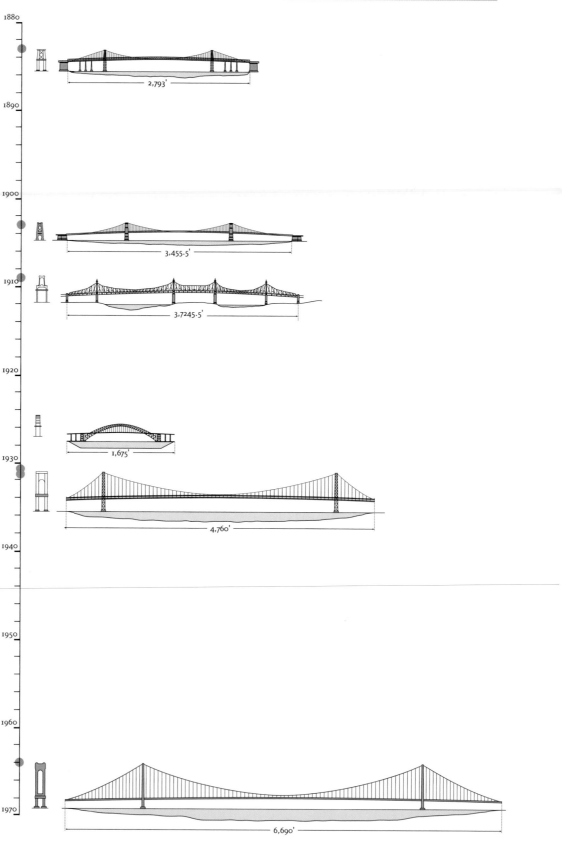

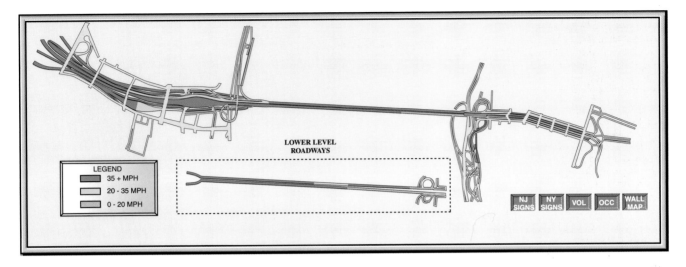

LOWER LEVEL
ROADWAYS

LEGEND
35 + MPH
20 - 35 MPH
0 - 20 MPH

NJ SIGNS · NY SIGNS · VOL · OCC · WALL MAP

Screen shot of George Washington Bridge and approach road speeds

The Naked Bridge

Although the George Washington Bridge is by no means the longest suspension bridge in the world, it remains among the most robust. It contains 113,000 tons of fabricated steel and 106,000 miles of steel cable wire—almost half the distance to the moon.

But the most unusual fact about the George, as she is affectionately called, is not her size or her strength; it is her appearance, which is something of a historical accident. The original design for the bridge saw her towers cloaked in masonry. But once partially built, the awesome skeletal beauty of the structure—as well as the cost of continuing construction—staved off the concrete cladding. By 1931, the Depression was in full force and the bridge's sponsor—the Port Authority—opted to leave the bridge half dressed, bringing construction in under the original $60 million cost.

Bridge Operations

Few New Yorkers appreciate how complex operating a bridge can be. Take the George Washington Bridge, for example. Its operations center collects information from

- **159 radar detectors,** which detect vehicle volume, occupancy of portions of the road, and speed;
- **39 cameras,** which provide a real-time look at the bridge's roadway conditions;
- **pavement sensors in the asphalt,** which show temperature, icing conditions, freeze point, and road surface information;
- **wind speed, air temperature, and visibility sensors,** which indicate dangerous conditions;
- **Highway Advisory Telephones and call boxes,** which provide highway and roadway information;
- **Variable Message Sign (VMS) sensors,** which show travel times but also indicate which of the signs' pixels and fans are (or are not) working.

In addition to staff at the bridge's Operations Center, 84 full-time and 23 part-time toll collectors are on hand to collect tolls from drivers who don't use E-ZPass. In total, it takes 300 people to keep the elegant bridge working.

Bridges & Tunnels

Bridge Maintenance   Just like pedestrian and railroad bridges, state law requires that all vehicular bridges be inspected every two years. Inspection is largely visual: technicians ride along the bridge undersides in motorized "travelers" looking for cracks, rusting, and corrosion or, alternatively, watch from the ground as trucks traverse the bridge. They may also strike the bridge with hammers, listening to vibrations from the concrete or steel. More sophisticated instruments—such as X-ray, laser, and acoustic devices—may be used to identify particular problems.

Inspection of a major bridge can involve as many as 50 people and take up to three months. Each inspection results in a bridge rating on a scale of 1 to 7; a rating of lower than 4 justifies inclusion in DOT's capital plan. A "flag system" is used to identify the existence of conditions that pose or could present a danger:

- **A red flag** is used to report the immediate failure of a critical structural component. Red flags must be addressed within six weeks. In 1988, for example, inspectors found the Williamsburg Bridge so crippled with rust that it was closed for two months for emergency repairs.
- **A yellow flag** is used to report a hazardous condition or the imminent failure of a noncritical component (one whose failure would not result in structural collapse).
- **A safety flag** is used to report a condition that presents a vehicle or pedestrian hazard, but that is not likely to result in any loss of reserve capacity or redundancy for the bridge. A missing railing or loose piece of concrete, for example, would warrant a safety flag.

Bridge Cleaning and Maintenance

Cleaning of expansion joints
Expansion joints, located at a bridge's surface level, are subject to a variety of elements: water, ozone, dust, and dirt, as well as to chemicals in salt products and gasoline. Preventing penetration of these is achieved by using compressed air and water to remove debris before cleaning and resealing the joints.

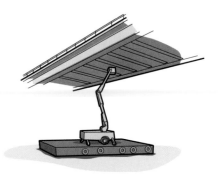

Bucket trucks *Mounted on a barge can be used in some cases to inspect the underside of a bridge.*

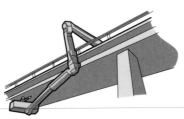

Bucket truck *To get a close look at a bridge's underside, maintenance crews use a specialized truck with an articulated arm that can curve underneath the structure.*

Debris removal *Debris can cause hazardous conditions on bridges; it also traps moisture and salt on the structure and can block proper drainage. Debris ranges from rocks to mufflers and wheel covers to paper, bottles, and cans.*

Paint removal and repainting
Paint removal is accomplished through abrasive blasting, but a containment area (including tarps, scaffolding, and cables) must be set up first. Three coats of lead-free paint are generally applied: the primer, intermediate, and top coats.

Cleaning of drainage systems
The cleaning of surface gratings, gutters, and downspouts calls for brooms, brushes, and a variety of other hand tools. Occasionally an air compressor will be needed to thoroughly empty certain gutters.

Paving Worn surfaces, which are usually made of a two-inch slab of bituminous concrete, are replaced and the roadways repaved.

Spot painting Surface contamination due to corrosive objects such as de-icing salts, bird excrement, or sea salt is generally removed by power washing with clean water or steam. Areas containing deteriorated paint are generally cleaned with hand tools.

Mechanical sweeper Mechanical sweepers move along the curb of the bridge deck to remove dust and debris.

De-icing De-icing trucks are used to spread abrasives and chemicals on road surfaces to prevent them from freezing over and causing unsafe driving conditions.

Bridges & Tunnels

Tunnels In contrast to New York's bridges, lauded in song and poetry, New York City's tunnels merit little attention. Yet the four vehicle tunnels that connect Manhattan with Long Island and New Jersey —the Brooklyn Battery, the Queens Midtown, Holland, and Lincoln tunnels—are a critical part of managing the flow of people into and out of the city every day.

The tunnels that run under the East and Hudson rivers were marvels of engineering in their day. Although the first trans-Hudson rail tunnel had opened as early as 1910, the larger size demanded by vehicular tunnels, coupled with the need to remove vehicle exhaust, presented more substantial challenges to builders. The system developed for the Holland Tunnel in the late 1920s—a two-duct system that relies on one duct to draw in fresh air and the other to suck out exhaust—would be adopted by vehicular tunnels worldwide and is still in operation today.

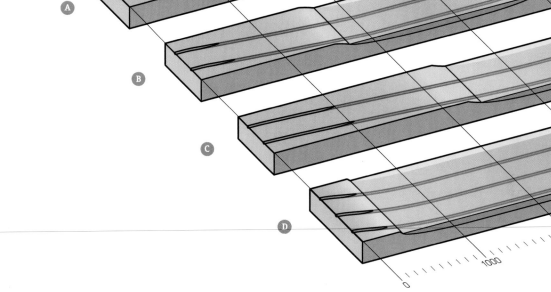

Tunnel Profiles

Length in feet

A The Brooklyn Battery Tunnel
When it opened in 1950, the Brooklyn Battery Tunnel was, at 9,117 feet, the longest continuous underwater vehicular tunnel in the world— a title it still holds. Two ventilation buildings in lower Manhattan, one in Brooklyn, and a fourth off Governors Island change the air in the tunnel every 90 minutes.

B The Queens Midtown Tunnel
The Queens Midtown Tunnel was opened in 1940 to relieve congestion on the city's East River bridges. Each of its tubes was designed one and a half feet wider than the Holland Tunnel in order to accommodate the wider cars of that period, and its maximum roadway gradient is 4%. The tunnel serves as the westbound terminus of the Long Island Expressway.

C The Holland Tunnel *The oldest of the region's vehicular tunnels, the Holland Tunnel connects Canal St. in Manhattan with 12th and 14th streets in Jersey City. Opened to great fanfare in 1927 as the first mechanically ventilated underwater vehicular tunnel, it was designated a National Historic Landmark in 1993.*

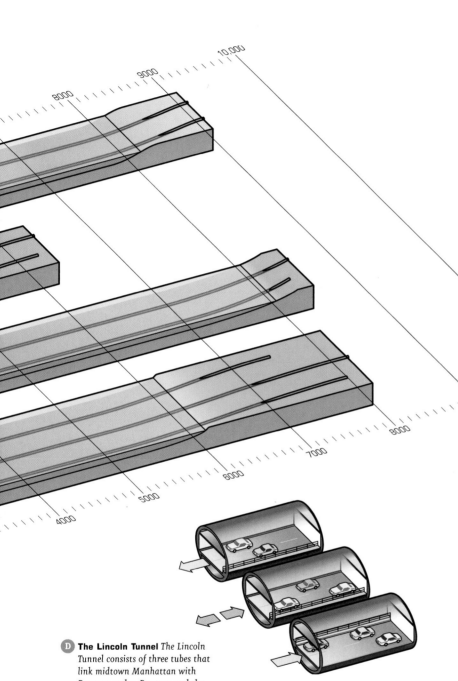

The Brooklyn Battery Bridge

Proposals for a tunnel between Battery Park and the Red Hook section of Brooklyn were approved by the city in 1930, but were put on hold with the onset of the Depression. Very keen to see the tunnel built, Mayor LaGuardia turned to Robert Moses—whose Triborough Bridge Authority was running a surplus—to fund, construct, and operate the new crossing.

Not nearly as passionate about tunnels as he was about bridges, Moses changed the plan for a six-lane tunnel to a six-lane bridge crossing; it would be cheaper to build and operate, carry more traffic, and almost certainly be more monumental. The new Brooklyn Battery Bridge would be a twin suspension bridge, held together by a central anchorage at Governors Island and connecting with the West Side Highway via a low-level causeway near Battery Park.

Notwithstanding the support of the City Planning Commission, Moses's bridge proposal ran into strong opposition both locally and in Washington. In July 1939, Franklin Roosevelt's secretary of war put an end to the project, claiming that the proposed bridge would be vulnerable to attack and would block access to the Brooklyn Navy Yard, giving life once again to the idea of a tunnel.

D The Lincoln Tunnel *The Lincoln Tunnel consists of three tubes that link midtown Manhattan with Routes 1 and 9, Route 3, and the New Jersey Turnpike in New Jersey. The three-tube configuration offers the flexibility to provide four lanes in one direction during rush hour or three lanes in each direction.*

Bridges & Tunnels

Inside the Holland Tunnel

Exhaust air moves through a duct on the tunnel ceiling. The air is changed every 90 minutes.

The tile finish requires constant cleaning. An estimated four million tiles were installed when the tunnel's ceiling was replaced in the late 1980s.

The tunnel's outer ring is composed of 14–15 cast iron sections—18" x 3'—bolted together. The seams are staggered for greater strength.

Ventilation buildings, the most visible parts of New York tunnels, are generally located on land on either side of the tunnel. The Holland Tunnel has four, which together contain 42 fans (28 of which are in operation at any one time) responsible for blowing fresh air into the tunnel through a duct below the roadway. Each fan is 80 feet in diameter.

Water moves out of the tunnel roadway through a system of curb drains. Most water leaks in the tunnel occur on land as a result of groundwater penetration or broken water mains.

The Exclusive Bus Lane (XBL)

Perhaps no aspect of regional traffic management is as unfamiliar to New Yorkers as the Exclusive Bus Lane (XBL), which provides access to the Lincoln Tunnel for city-bound commuters from New Jersey. The XBL is actually a 2.5-mile-long stretch of a westbound lane, from the tunnel to the New Jersey Turnpike, converted to eastbound use during weekday morning peak periods. Over 100 different bus lines, numbering 1,700 or so buses, use the lane each day. An estimated 60,000 commuters save somewhere between 15–20 minutes in travel time, compared with the normal congested Route 495 approach to the tunnel.

Tunnel Maintenance New York State law does not require specific repair or maintenance procedures for tunnels. However, tunnels coming into the city are regularly inspected for structural integrity by either the MTA or the Port Authority. In the case of the Port Authority, comprehensive tunnel inspections are undertaken by outside consultants every two years to ensure structural integrity and as a supplement to annual routine inspections. Deficiencies are generally classified as high priority, priority, routine priority, or routine—and addressed accordingly.

To inspect tunnels, inspectors crawl through the system looking for cracks or missing chunks of concrete, broken bolts, and water seepage or leaks. Inspectors will tap the concrete with a hammer, called "sounding," and listen to the echo. Inspections go beyond the tunnel itself to include ventilation and service buildings as well as ancillary structures such as retaining walls and pump rooms.

In addition to inspections, the tunnels require a sizable retinue of maintenance staff who carry out daily cleaning, repair, and maintenance work. The Holland Tunnel, for example, maintains a staff of 83; the Lincoln requires 79. Personnel service costs for each of these tunnels on an annual basis exceed $4 million.

Tunnels need to be cleaned regularly; the Holland Tunnel, with one of the lowest clearances, will turn black from truck exhaust in just two days. Giant electrical toothbrush trucks, sporting arms with rotating brushes, move through the tunnel three to four times a week. They rely primarily on water for cleaning, to avoid any environmental concerns.

MOVING FREIGHT

Cities are great consumers of goods, and nowhere is that more true than New York. Each day, hundreds of millions of dollars' worth of goods moves into the region by ship, rail, air, or truck—much of it destined for the city. Though largely taken for granted, trade is as important to life in New York today as it was when the city was founded.

Yet the physical manifestations of trade are now all but invisible to most New Yorkers; gone are the docks and wharves pulsing with exotic cargos. Today, trade moves much more quietly—and much more efficiently— through the region's port and airport facilities, through half a dozen rail freight yards, and through a handful of wholesale markets. The occasional sight of a cargo ship steaming into port or a freight train running down the Hudson belies a complex system of transportation logistics that underpins the commercial life of the city.

another, the "merchandise trains" destined for New York contain a mix of commodities coming from numerous producers and earmarked for dozens of consumers.

These freight trains—some up to 120 cars in length—are environmentally friendly. On average, each train replaces 280 trucks that would otherwise be making the same journey. Unfortunately there are not more of them: rail cargo today makes up only about 5.6 percent of the freight moving through the region, down from a high of roughly 40 percent in the early 1940s. The drop in rail traffic is in part a reflection of the region's appetite for imports, but also a function of increased competition from long-distance trucking industries.

The fact that rail has survived at all is due in part to a largely successful intervention on the part of the federal government over 25 years ago. Once the primary means of moving goods from the west, rail service had deteriorated so badly by 1976 that Washington, D.C., stepped in to create the Consolidated Rail Corporation (Conrail) out of the bankrupt Penn Central Railroad and five other struggling lines in the Northeast. Roughly $7 billion of taxpayers' money was invested in trains and track repair, and in 1987—after some reasonable success—Conrail was sold to the public. With a monopoly on freight traffic into and out of the metropolitan region, the company proved an attractive target for both Norfolk Southern and CSX railroads and the two railroads jointly purchased Conrail in 1999 for $10.3 billion. Although Conrail continues to exist today as a subsidiary of both companies performing switching services at local yards, most Conrail assets were divided up between the two railroads.

Each week, roughly 1,750 railcars move through the metropolitan region— a small but important

part of the region's freight lifeline. Each of these railcars represents a shipment loaded hundreds or thousands of miles away and destined for consumers in the New York area. Unlike "unit trains" made up of a single commodity moving from one place to

Rail Freight

Reinventing the High Line

One of the most talked-about relics of New York's rail freight era is Manhattan's High Line, a 1.5-mile-long, 30-foot-wide elevated rail deck running from West 34th St. to Gansevoort St. Now due to become a linear park, it was built in the 1930s as an attempt to reduce congestion on 10th Ave. For 30 years, the High Line brought food and merchandise into Manhattan—until improvements in highways led to a falloff in rail freight in the early 1960s.

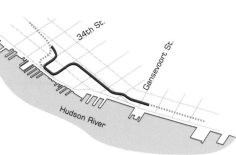

Little Ferry Yard

River Line

Hudson Line

Oak Point Link

Bergen County Line

North Bergen Yard

Oak Point Yard

Harlem River Yard

River Line

Kearny Yard

Croxton Yard

P & H Line

Fremont Secondary Line

Montauk Branch

Lehigh Mainline

Oak Island Yard

National Docks Secondary

Fresh Pond Yard

LIRR Main Line

Greenville Branch

Raritan Valley Line

Greenville Yard

E-Rail Terminal

Union County Central Railroad

South Brooklyn Marine Terminal

Arlington Yard

Chemical Coast Line

Travis Line

65th Street Yard

Bay Ridge Branch

Today, the New York–New Jersey region supports more than a dozen rail terminals, served by three major railroads: the Canadian Pacific, handling traffic to and from eastern Canada, and CSX and Norfolk Southern, both of whose routes lie primarily to the south and Midwest. They are supported by seven smaller regional or terminal railroads: the New York and Atlantic, serving Long Island's freight customers; Express Rail, serving port users; the New York Cross-Harbor Railroad, serving businesses requiring carfloat service; the South Brooklyn Railway, serving NYC Transit's needs; the Providence & Worcester Railroad, running from the region to parts of Connecticut, Rhode Island, and Massachusetts; the New York Susquehanna and Western, serving traffic between upstate New York and New England and the mid-Atlantic region via a connecting line around New York City; and the Port Jersey Railroad, providing local switching services in northern New Jersey.

Rail Freight

Railcars There are roughly 1.3 million railcars in the United States, of varying shapes and sizes. Many are highly specialized, designed to carry lumber, chemicals, forest products, or autos, for example. Others are variations on standard "hopper" or "gondola" cars. Hopper cars generally handle dry bulk commodities impervious to weather conditions—stone, gravel, or coal, for example; gondola cars, either covered or open, are used to ship heavy or bulk products such as scrap metal, steel, wood chips, and aggregates. Refrigerated cars, with diesel-powered cooling units, are used to meet the long-distance travel requirements of fresh or frozen products.

Regardless of type, all railcars carry markings on their flanks. Generally these will include the car number, the railroad trademark or logo, and the name or initials of the car's owner. They will also carry abbreviations referring to their cubic and weight capacity, length, width, height, and date built.

Boxcar *A boxcar is a fully enclosed car used to transport commodities.*

Gondola car *A gondola car is a low freight car with a flat bottom, fixed sides, and no roof. It generally carries bulk goods such as stone or steel.*

Tank car *Tank cars are used to transport liquids, compressed or liquefied gases, or solids that are liquefied before unloading.*

Covered hopper car *Covered hopper cars are used for handling bulk commodities that can't get wet. They have openings for either top or side loading.*

Trilevel auto car *A trilevel auto car consists of a three-level steel rack that holds 12 standard sedans or 15 compact cars.*

Refrigerator car *Also known as a "reefer," a refrigerated car is used to move goods needing refrigeration. Before the era of gas-powered coolers, these cars were loaded with ice.*

Floats No More Prior to the opening of the Holland Tunnel in 1927, nearly all domestic freight destined for New York terminated its rail journey in New Jersey. From there, it crossed the river on cargo ferries or on carfloats, barges specially outfitted with rail tracks for cargo moving from one rail system to another. At one time, dozens of carfloat bridges existed along Brooklyn's shoreline.

Today, in an age of multiple truck routes across New York Harbor, only one carfloat operation remains. Known as the New York Cross-Harbor Railroad, it moves roughly 2,000 cars each year between South Brooklyn and Greenville Yards in Jersey City—about the same amount that was handled *each day* in 1965. Between 15 and 20 railcars can be ramped from the shore onto tracks on a waiting barge, which is hauled by a tug across the river to an unloading yard.

Most cargo, primarily for Long Island customers, is of the "not in a rush" variety: the journey takes longer but is cheaper than the alternative rail journey up the Hudson and across the river at Selkirk, near Albany.

Center beam bulkhead flatcar
These cars are often used to transport lumber or sheets of drywall, which are stacked on either side of a center beam.

Special-purpose depressed center flatcar *These flatcars are generally used to haul extremely oversized items.*

RoadRailer *The RoadRailer is a specialized trailer vehicle designed to move over the highway, but also to be pulled in a train. Originally designed in 1952, it is used by a number of railroads, including Amtrak for mail service.*

COFC *Container-on-flatcar (COFC) service is a common sight on freight trains rumbling down both sides of the Hudson River.*

Double-stack *Double-stacked containers are placed on specially designed low-level chassis to meet the 22-feet clearances common on rail lines in parts of the United States.*

TOFC *Trailer-on-flatcar (TOFC) service consists of truck trailers riding on flatcars. This system is generally referred to as "piggybacking."*

Intermodal Cargo The lion's share of rail cargo into and out of the New York region is what's known as "intermodal" cargo; i.e., cargo moving by more than one means of transportation. Truck trailers, for instance, may be loaded on rail flatcars and taken long distances by train. Containers may travel the same way, moving easily between any combination of ship, rail, and truck. At least one specialized vehicle—the RoadRailer—can move on both road and rail.

Increasingly, however, intermodal traffic moves in containers. Thanks to special low-level rail chassis, containers can be stacked on top of each other for long journeys across the country. Known as "double-stack" trains, this mode of rail travel has proved so efficient that Asian cargo headed for the New York region is often dropped off on the West Coast and completes its journey to the East Coast this way—in what has become known as the "mini-landbridge" system. Double-stack trains are also loaded here at the port, and in other northern New Jersey terminals, with imports bound for eastern Canada and the Midwest as well as exports.

What's in a train? As anyone who has ever waited at a grade crossing for a freight train to pass knows, these trains can be very long indeed. And that is precisely the economics that underpin rail freight: link as many different cars as possible going roughly to the same location, pulled by the same locomotive. The unusual mix of commodities that can result is represented here by the lineup of a train delivered to the New York & Atlantic Railway on August 26, 2003.

Wine   Pulpboard

Rail Freight

Classification Yards Railroads' advantage over trucking turns on their ability to move many diverse shipments over a relatively long distance on one train. But making up a train is a cumbersome process, as no two cars may have the same origin and destination. Here's where the classification yard comes in: cars are collected from shippers and assembled into trains for travel to a second yard, where they are broken up and sorted for delivery to customers.

Two kinds of classification yards predominate—flat yards and hump yards. Flat yards consist of a set of parallel tracks interconnected by switches, and rely on switch engines to move cars in blocks or individually. Hump yards, in contrast, are characterized by a track raised above the rest of the yard; the switch engine pushes the car over the hump and gravity accelerates it onto its predetermined track. Automatically operated retarders brake each car's wheels so that it couples at just the right speed to the cars already lined up on the track.

Within the New York–New Jersey region, several classification yards act as part of the regional rail network. Oak Island, just north of Port Newark, New Jersey, is operated by Norfolk Southern and CSX. Oak Point Yard, in the Bronx, is the largest classification yard within New York City. It serves as a classification and staging yard for freight rail traffic to and from Long Island over the Hell Gate Bridge. Most traffic bound for New York City from the west moves over Selkirk Yard, just south of Albany.

Selkirk Yard Nearly all freight moving directly into New York City comes through the Selkirk Yard, located eight miles south of Albany and operated by CSX. With 70 tracks— the longest will hold 70 cars and the shortest 37—Selkirk is the largest classification yard on the East Coast.

Receiving yard *After the engine is removed from the train, the remaining cars are inspected for mechanical defects.*

Engine house *The engine house is where the engines are inspected and maintenance and repairs are undertaken. It is often located at the center of the yard, so as not to disrupt the humping activity.*

Pulpboard —————————————————————————————— Propane —————— Corn Feed

For Local Delivery Perhaps the closest thing New York has to a local railroad is the New York & Atlantic (NY&A), which operates a 269-mile system that primarily serves customers on Long Island. It is a recent phenomenon: until 1997, the rail freight business on Long Island (including Queens and Brooklyn) had been the purview of the Long Island Rail Road or its predecessors. At that time, NY&A, a subsidiary of the Anacostia and Pacific Railroad holding company, was awarded a 20-year concession from the MTA with the goal of reversing a 25-year decline in Long Island rail freight volume. Employing only 30 people, it runs eight trains a day (six days a week) along tracks it shares with the commuter railroad. Its 18,000 annual carloads consist primarily of aggregates, scrap paper and metal, forest products, chemicals, and food products.

Car repair yard *The car repair yard is the location of light repair. Maintenance men will be dispatched throughout the yard, generally to tracks designated by type: hopper, gondola, etc.*

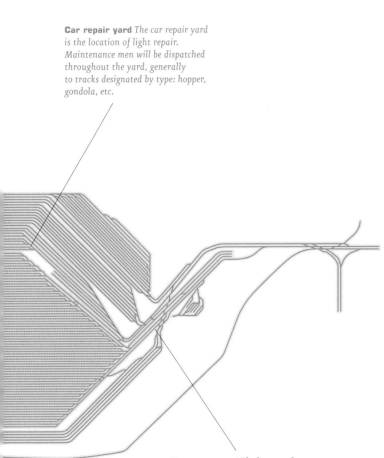

Departure yard *Blocks move from the classification yard to the departure yard, where they are made up into trains. Car inspectors look over the train, attach the air hoses, and couple the engines.*

Pulpboard   **Rice**   **Flour, Bagged**

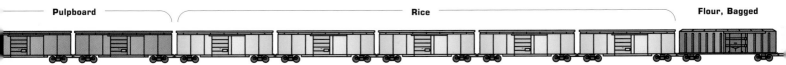

Rail Freight

How a Hump Yard Works

Hump yards are most efficiently used to classify trains made up of cars going to many different destinations. The hump locomotive travels only one train length in order to "classify" or segregate the entire train.

Pit *As a train leaves a receiving yard and moves toward the hump that signals its entrance into the classification yard, the train rolls over a glassed-in pit under the track. An inspector will examine the couplers, gears, and brakes among other things.*

Hump *The hump is an artificial hill, generally about 20–30 feet high. A hump engine is attached to the train and pushes the cars over the hump at a rate of 3–4 per minute, down a grade of between 2% and 4%. Some humps include a scale and inspection pit.*

"Do Not Hump" *Occasionally freight cars rumbling through the New York region carry the admonition "do not hump." These are generally cars carrying fragile or high-value loads—liquor, bricks, glass, or delicate food products, for example. Sending these trains "over the hump" could damage the goods—hence the warning.*

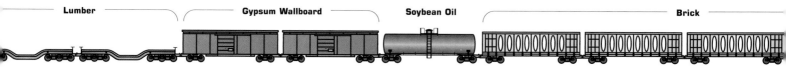

Lumber   Gypsum Wallboard   Soybean Oil   Brick

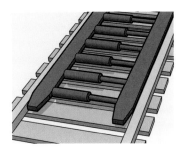

Retarder *Speed meters monitor the cars' speed once over the hump, and this information is passed to the retarder operator. This indicates just how much to slow the train (via mechanical retarders that grip each car's wheel flanges) so that coupling speed does not exceed 4 mph.*

Classification yard *Once over the hump, cars roll into the classification yard, where each track is assigned a destination. As cars accumulate in the yard, blocks of similarly destined cars are built. Yards are often called "bowls," as most track slopes to the center.*

Building a train *Power-operated switches control the track routes from the classification yard to the pullout leads. One or more leads feed the departure yard or yards, where trains are made up based on destination.*

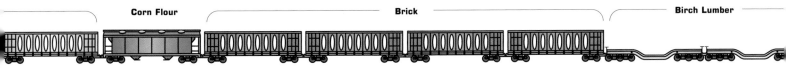

Corn Flour · Brick · Birch Lumber

Rail Freight

Transcontinental Freight

Not all that long ago, produce moving from the West Coast to New York was shipped almost exclusively by rail: potatoes and onions from the Northwest, fruit from California, for example. However, as trucking became more competitive and efficient (in part due to the interstate highway system) and rail service deteriorated, the tables began to turn. Today, only a minority of produce moves into the New York region by rail: Sunkist, for example, used to load more railcars in a day than it does now in a year.

But transcontinental rail freight is trying to make a comeback, with clean, refrigerated cars ("reefers") and—most important—more reliable train scheduling. Though cross-country rail transit times are roughly double those by road, the cost of moving by reefer is roughly half that of truck thanks to labor shortages, rising fuels costs, and increasing highway congestion. As a result, rail is making inroads with the more durable types of western produce: carrots, onions, celery, potatoes, broccoli, and citrus fruit.

The Journey of a Carrot

1. At a packing shed near Bakersfield, California, workers for Grimmway—the world's largest carrot grower— hand-load over 2,500 50-pound bags of carrots into a refrigerated railcar about to begin its eight-day journey to the Bronx.

2. Once packed, a local train collects the car and brings it to a classification yard in Fresno. Along with refrigerated trains collected elsewhere and several cars carrying nonperishables, it leaves for the Union Pacific's Northern California hub, Roseville Yard, where it is coupled with another block of refrigerated cars.

3. At Roseville, engines are added and the train, limited to 85 cars to enable it to manage the steep grade of the Sierra Nevada, departs for Bailey Yard in North Platte, Nebraska —the largest classification yard in the world. At North Platte, the refrigerated cars will be combined with reefer loads from Idaho and Oregon and sorted into two new trains —one headed for Selkirk, New York, and the other for Waycross, Georgia.

4. The Selkirk-bound train arrives at Union Pacific's Proviso Yard west of Chicago and is then moved by a local switching railroad to a CSX yard nearby. The Union Pacific crew is replaced by a CSX crew, although the Union Pacific's engine will continue on the train throughout its journey to Selkirk.

Plastic Pellets   **Plywood**   **Potatoes, Fresh**   **Oak Flooring**   **Beer**

"Big Milk" To this day, rail freight buffs remember with fondness the "milk trains" —the trains that brought nearly a million gallons of milk each day into the region from farms in upstate New York. These trains carried cans of milk in special insulated cars (packed with ice in the summer) to "milk yards" in the region. Three terminated at the 60th St. Milkyard on the west side of Manhattan, dropping off cars at the Bronx Terminal Market and the 130th St. Milkyard in Manhattan en route; a fourth terminated in Weehawken, New Jersey. The trains generally began their rural runs in the afternoon, arriving in the city during the wee hours of the night.

Tank cars for milk replaced cans around the turn of the last century: precooled milk was pumped into the 6,000-gallon car at the origin of the journey and pumped out into waiting trucks at the receiving end, where it was moved to pasteurization plants. Though the tank cars brought greater levels of efficiency than the can cars, they proved no match for the truck. By the 1950s, trucks traveling on improved state highways offered a more direct and faster haul from the country milk station to distribution points in the city and "Big Milk"'s days were numbered.

5. *The train arrives at a hump yard in Selkirk, where it is sent "over the hump" to become part of a train headed south along the east side of the Hudson to Oak Point Yard in the Bronx. This train only runs at night, as it shares track with Metro-North and must be clear of Croton Harmon by 4:30 a.m., when the morning commuter rush begins.*

6. *The carrot car, along with 19 other reefers, arrives at Oak Point Yard and is moved alongside a shed at nearby Hunts Point Market, the largest wholesale produce complex in the nation.*

7. *Here, the car is unloaded and its contents distributed to the ultimate consignees. The car will return to the West Coast either empty or loaded with westbound cargo.*

Pulpboard   Onions, Fresh   Corn Starch   Lumber

By almost any account, New York owes its origins as a commercial center to its advantageous location on maritime trade lanes. With one of the world's great natural harbors at its front door and a mighty river at its back, maritime trade gave rise to the young city in the eighteenth century and propelled it to national prominence in the nineteenth. The opening of the Erie Canal in 1825 served to cement its commercial position and by 1860 nearly half of the nation's trade moved through the Port of New York.

For a century or so, Manhattan was the epicenter of this trade. Initially, it was the docks and wharves of South Street that bustled with mail and other cargo ships. Businesses sprang up along the piers to serve the trade, and longshoremen settled their families in adjacent neighborhoods. Soon, the network of shipping activity—and the intricate web of finger piers it required—spread to Manhattan's West Side, as well as to the Brooklyn, Hoboken, and Jersey City waterfronts.

These finger piers served as the lifeblood of the city and—in times of war—the nation. During World War II, there were roughly 750 active piers in the port—able to berth 425 oceangoing vessels simultaneously. Within two generations, however, nearly all of them disappeared. No event was more responsible for that transformation than the invention of the container in the 1950s, which offered tremendous efficiencies and greatly expanded maritime trade. Its demands for large tracts of open space found an outlet in the swampy backwaters of New Jersey, and within a generation the din of the working waterfront was but a memory to most New Yorkers.

Maritime Freight

The West Side waterfront, circa 1869.

In the years since then, New York's maritime trade has grown dramatically—more than 80 million metric tons of cargo move through the port each year—but it is also less visible. Much of it moves over docks at the Port Newark/Elizabeth Marine Terminal in New Jersey, whose 2,100 acres sit just east of Newark Airport. Supplemented by New York Container Terminal on Staten Island, Red Hook Container Terminal in Brooklyn, and a number of private marine terminals in New Jersey, the little-heralded complex serves in many ways as the economic lifeline of the region.

The Harbor Today, New York Harbor remains among the city's greatest assets. It covers 650 miles of shoreline, reaching from the banks of Sandy Hook in New Jersey around Staten Island and northward along the contours of Newark Bay and the Hudson and East rivers. Although it is a natural harbor, it is not a naturally deep one—silty deposits from the Hudson, Hackensack, and Passaic rivers give it a natural depth of between 18 and 21 feet. As a result, harbor traffic must stick carefully to a predetermined set of man-made and well-maintained navigation channels and anchorages.

Passaic River

Global Marine Terminal *is a private container terminal located in Bayonne, New Jersey.*

Hudson River

East River

Port Newark/Elizabeth, *the destination for much of the region's cargo, is accessed via the narrow Kill van Kull located on the north side of Staten Island.*

Brooklyn Marine/Red Hook Container Terminal *is located opposite Governors Island in the Red Hook section of Brooklyn.*

Hackensack River

Newark Bay

The Statue of Liberty *is the official center of the port, with the "port district" radiating out 25 miles in each direction.*

New York Bay

Kill van Kull

New York Container Terminal, *formerly known as Howland Hook Marine Terminal, is on the north shore of Staten Island and handles both containerized and non-containerized traffic.*

Ambrose Channel

Arthur Kill

Sandy Hook Channel

Maritime Freight

Entering the Harbor Each year, more than 12,000 ships enter or leave New York Harbor. Roughly 40 percent of them are tankers or "drugstore ships," carrying refined products or crude oil. Another 45 percent are laden with containers, destined for warehouses and distribution centers in the region. The rest are bulk or break-bulk (cargo consolidated into smaller, noncontainerized units) vessels, often carrying single products such as iron, steel, or forest products. With the exception of tankers headed for terminals along the Arthur Kill, nearly all of them pass under the Verrazano Narrows Bridge, the unofficial gateway to New York Harbor.

The passage under the Verrazano is but one of several tricky maneuvers today's cargo vessels must make before reaching port. In some places, the harbor bottom is made up of soft material like sand, silt, or clay; in other places, the seabed is rock—a less forgiving matter. The combination of sharp turns, wild currents, and a preponderance of reefs and shoals means that only a trained harbor pilot is allowed to guide large ships into port: even the most experienced ship captain must relinquish the wheel to a Sandy Hook pilot when his or her ship reaches New York waters.

Making for Port

Staten Island

The captain of the ship cedes control to the Sandy Hook pilot, who takes the ship through the Narrows and into the harbor.

A skiff from the pilot boat moored offshore approaches the ship, and a Sandy Hook pilot climbs up a rope ladder to board the vessel entering the harbor.

Ships approach from one of the three major shipping lanes—Barnegat (from the south), Hudson (from the east), or Nantucket (from the north).

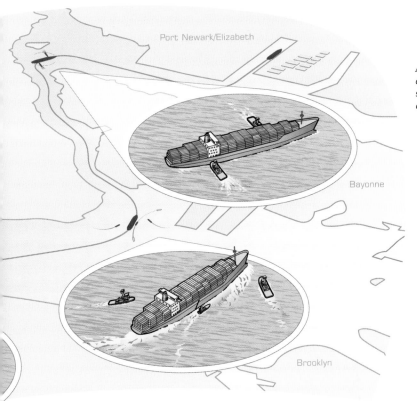

Port Newark/Elizabeth

Bayonne

Brooklyn

A docking pilot takes over as the ship approaches Port Newark. He or she will guide the ship into her berth at the port.

Tugs come alongside the ship, to assist in making the sharp turns necessary to travel through the Kill van Kull between Bayonne and Staten Island.

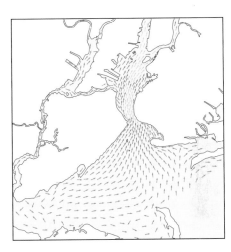

Monitoring Waves and Tides
Current velocities and depths in New York Harbor vary widely, across the harbor and over the course of a day. The National Ocean Service, part of the Department of Commerce, maintains a model to provide mariners with information that can help them time their journey or decide which route to take. It is three dimensional and relies on real-time wind and water-level data to predict water levels and currents at thousands of locations across the harbor.

→ < 0.3 knots
→ 0.3–0.6 knots
→ 0.6–1.0 knots
→ 1.0–1.3 knots
→ > 1.3 knots

The Invisible Pilots Since 1694, when the Colonial Assembly commissioned a group of local sea captains to aid ship masters entering New York Harbor, pilots have been responsible for navigating ships through New York Harbor's treacherous waters. Relying initially on oars and sail, local pilot groups in New York and New Jersey competed for the patronage of the incoming vessels. A tragic accident in 1888 forced New York State to act to combine local pilot companies; seven years later, the New York and New Jersey pilots groups merged and the Sandy Hook Pilots Association was born.

Today, a full century later, the Sandy Hook Pilots still have a monopoly on this business. Some 76 pilots take turns manning one of two large pilot boats stationed around the clock off Sandy Hook, assisting on average 35–40 incoming or outgoing ships each day. And they are well trained: a seven-year apprenticeship must be followed by seven more years of work as a deputy pilot.

Maritime Freight

Managing Harbor Traffic

The U.S. Coast Guard, once part of the Department of Transportation and now part of the Department of Homeland Security, is responsible for monitoring and coordinating New York's harbor traffic. It does this largely through its 24-hour Vessel Traffic Service (VTS), based at Fort Wadsworth on Staten Island. Staffed by a mix of civilian and military personnel, the service gathers and disseminates real-time information about marine movements via three radio frequencies:

- **Channel 11** is provided for initial check-in, when a boat is getting under way from a mooring or entering the harbor.

- **Channel 12** serves the Arthur Kill and East River and is used by the Coast Guard to administer the harbor's anchorages.
- **Channel 14** covers boats steaming through the main shipping channel, including the Lower and Upper Bay, the Kill van Kull, Newark and Raritan bays, and Sandy Hook Channel.

In addition to coordinating vessel movements, the Coast Guard monitors and administers boat "parking" in the harbor at the three federally designated anchorages at Bay Ridge and Gravesend, off Brooklyn, and at Stapleton, off Staten Island. Vessels are required to provide four-day advance notice before arriving at an anchorage and are permitted to stay for a limited period of time—generally 30 days.

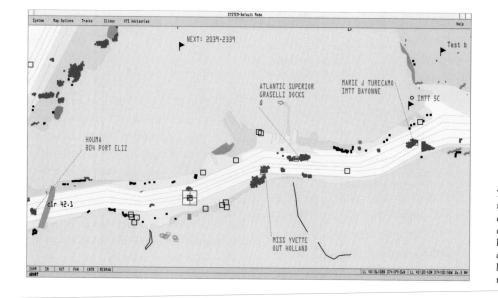

The Coast Guard's Vessel Traffic Service monitors the location and destination of a variety of harbor craft around the clock. Vessels labeled in green on this screen shot are carrying general cargo; vessels labeled in red are carrying hazardous materials, generally petroleum.

Monthly Volumes of Intra-Harbor Traffic

1 ship = 100 vessels

| Tank ships | Freight/other cargo ships | Tugboats | Passenger and ferry boats | Public vessels |

Vessel Types The fleet of cargo ships that ply the harbor's waters are largely unrecognizable to the untrained eye. Though they can commonly be seen at anchorage or steaming purposefully up the Hudson to Albany, few New Yorkers can differentiate one cargo ship from another. And yet nowhere is the adage "form follows function" more true than in commercial ship design, where the contours of the ship are designed to maximize carrying capacity and facilitate loading and unloading.

Workhorse of the Harbor

Even landlubbers know that tugboats are the invisible workhorses behind freight movement in the harbor, and the sight of powerful tugs towing or nudging inert barges of bulk commodities up the Hudson or East rivers is a fairly common one. But few New Yorkers realize that even the newest and fastest cargo ships rely on tugs to help them negotiate the sharp turns that form part of the harbor's main shipping channels.

For well over a century, two Irish families have dominated the tug business in New York—the Morans and the McAllisters. Founded as small family businesses in the latter part of the nineteenth century, both are now dominant players in towing and transport up and down the East Coast. Notwithstanding their geographic expansion, they remain fixtures in New York Harbor: today 16 Moran tugs and 20 McAllister tugs—ranging from 1,750 hp to 6,300 hp—work full time in the harbor.

Reefers *Refrigerated cargo moves in ships known as "reefers," which provide climate-controlled conditions throughout a commodity's journey. Perishable commodities like meat and fruit typically move in them.*

Dredgers *A variety of types and sizes of dredge boats are in New York Harbor at any given point in time, involved either in maintenance dredging or harbor-deepening work.*

Container ship *Container ships originally carried a few dozen boxes on the deck of a freighter or converted tanker. Today, they are purpose-built and carry as many as 5,000 containers above and below the deck.*

Car carrier *Car carriers are specially designed to maximize the number of cars that can be stowed within their walls. Doors open and exit ramps are designed to allow longshoremen to drive the new cars off the ship as quickly as possible.*

Tankers *Tankers are a familiar sight in New York Harbor, either coming from or heading to the area's oil storage facilities.*

Bulker *Bulk ships handle a variety of bulk commodities, from bananas to forest products and paper to coffee.*

Maritime Freight

Dredging Basics

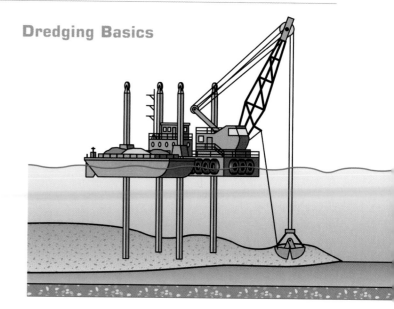

In areas of relatively soft material, a clamshell bucket attached to a barge-mounted crane removes the sediment and places it on an adjacent scow, which will be towed to the ultimate disposal site.

Harbor Maintenance

Left to its own devices, New York's harbor would silt up quickly, driving cargo and cruise ships to other ports. To prevent this from happening and to maintain safe channel depths for shipping, dredging is undertaken year-round throughout the harbor. On average, some three million cubic yards of material are removed from the bottom of the harbor each year by the U.S. Army Corps of Engineers. Much of this is mud or sand; the remainder is made up of clay, rock, or glacial till.

Harbor maintenance has been an issue for New York since the middle of the nineteenth century. As clipper ships gave way to steamships, and ultimately to cargo ships and ocean liners that drew even more water, deeper channels were a necessity. Ambrose Channel, the entrance to New York Harbor, was first taken down to 30 feet in 1884, initiating a program of regular deepening there and in the most heavily used shipping channels throughout the harbor.

Today, that deepening work continues, and an ambitious project is under way to take 40- and 45-foot-deep stretches along the Ambrose, Anchorage, Port Jersey, Arthur Kill, and Kill van Kull channels down to 50 feet. This deepening project is enormous in scale: it involves up to 80 pieces of dredging machinery—the largest concentration of dredging equipment ever assembled in this country. At a total cost of nearly over two billion dollars, it is a very expensive process and one funded in part by federal dollars. It is also a complex one: much of the sea floor being deepened consists of rock that must be blasted before it can be removed.

This deepening project alone involves the removal of an estimated 53 million cubic yards of sediment. Whereas historically this material would have routinely been placed at the "Mud Dump" south of Sandy Hook (renamed the Historic Area Remediation Site, or HARS), today it is carefully tested for contaminants and put to more beneficial use. The most contaminated sediment is mixed with cement or fly ash and used for landfill cover, while the cleaner sediment is earmarked for shoreline stabilization or fishing reefs, or used to cap contaminated underwater dump sites.

Harbor Depths Over Time

Depth in feet

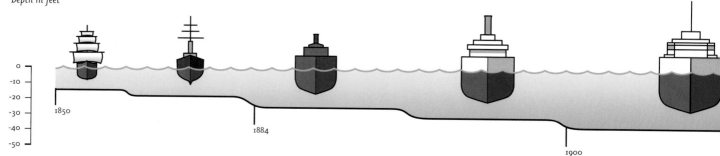

0
-10
-20
-30
-40
-50

1850

1884

1900

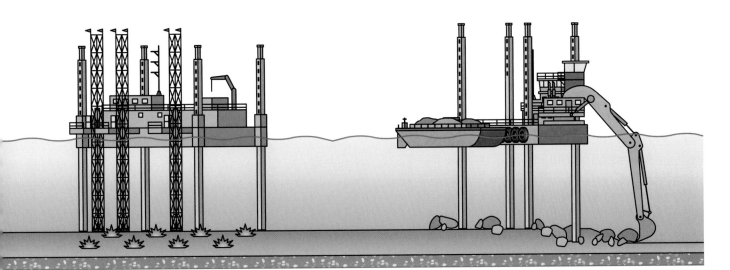

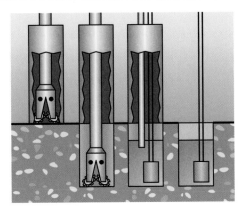

Where rock is present, a drill boat is used. Holes are drilled six feet below the required depth, roughly 10 feet apart, and then filled with Porvex, a liquid explosive. Once the blast has occurred, an excavator dredge is used to collect the loose rock and a survey boat will be brought in to certify the new depth of the channel.

Among other equipment, New York is the current home of "T-Rex"—the biggest barge-mounted dredge in the world. Its bucket holds 13 cubic yards, making it roughly the size of a garbage truck, and it can dig down to 65 feet. A screen appears in the cab of the dredger to pinpoint the depth of specific areas at and around the dredging site.

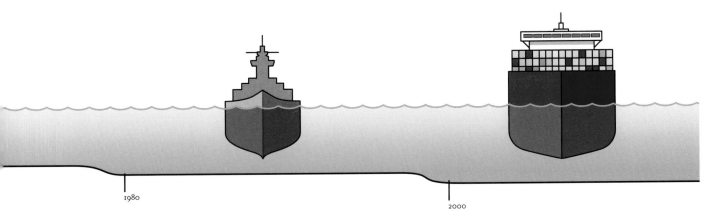

1980

2000

Maritime Freight

The Port Though New York is no longer the nation's dominant trade gateway, its port remains the third largest in the country (after Los Angeles and Long Beach) by tonnage, accounting for about 12 percent of the nation's maritime trade. The lion's share of this moves through the Port Authority's Port Newark/Elizabeth Marine Terminal on Newark Bay. Now fifty years old, the port complex there bears only a faint resemblance to its appearance at the dawn of the container age, when developing a port on 2,000 acres of New Jersey swamp was considered a very speculative proposition. Few would have predicted just how quickly or how completely this once remote backwater would be transformed into the massive international trade nexus it is today.

At the heart of the port's operation are containers. Nearly 1,200 acres of land are dedicated to moving them on and off ships, putting them onto railcars, or storing them for movement onto a ship or waiting truck. In addition to containers, the port is also home to two auto preparation centers, a bulk liquid–handling facility for edible fats and oils, two orange juice concentrate blending facilities, and bulk cargo–handling centers for gypsum, cement, scrap metal, and salt. It also features on-dock rail facilities, which handle the small but rapidly growing segment of port traffic moving to or from inland points by rail.

Mapping Maritime Trade *China is the port's largest trading partner, accounting for almost 20 percent of port activity. Italy, Germany, Brazil and India are the next largest trade markets. Not surprisingly, as a population center, the port imports more than it exports—seven times as much. In total, the port's trade was valued at $113 billion in 2004.*

Top Containerized Trade Partners *in 20-Foot Equivalent Units (TEUs)*

Export Import

———	———	< 32,500 TEUs
———	———	< 75,000 TEUs
———	———	< 150,000 TEUs
———	———	< 300,000 TEUs

UK
Netherlands
Belgium
Germany
Spain
Italy
New York
India
PRC
Republic of Korea
Hong Kong
Indonesia
Brazil

An Overview of the Port

Warehousing and distribution The port provides warehousing and distribution areas, including refrigerated warehousing for food products.

Orange juice facilities The port has two separate orange-juice blending facilities.

Manufacturing facilities Several manufacturing facilities are located right at the port, including a copper wire production plant and a wallboard plant.

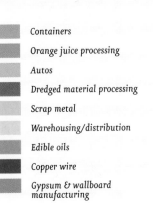

Autos The tristate area is the largest auto-handling market in the United States and car carriers are regular visitors to the port. The total volume of vehicles handled through the port in 2004 was 728,720, including small trucks, vans, SUVs, and automobiles.

On-dock rail Cargo increasingly moves directly from ship to rail, thanks to the introduction of what is known as "on-dock" rail facilities at the Elizabeth Marine Terminal. Nearly 300,000 containers move out by rail from the port each year.

Containers The largest amount of acreage at the port is devoted to container terminals. Over four million TEUs (20-foot equivalent units) move through the port region each year, many of them at Port Newark/Elizabeth.

Containers

Orange juice processing

Autos

Dredged material processing

Scrap metal

Warehousing/distribution

Edible oils

Copper wire

Gypsum & wallboard manufacturing

Maritime Freight

The container capacity is the total "cube," or cubic measurement, a container can hold. Capacity, or internal volume, is determined by multiplying the internal dimensions of the container, including its length, width, and height.

Each container has an identification code, the container number. It combines a four-letter character that identifies the owner and a seven-number character that identifies the container. The number can be found on the outer and inner side of the container.

In order to prevent their sliding into the ocean, containers are secured to each other by a system of interlocking bolts which are then attached to cross-tie lashings.

Containers are made of corrugated metal, generally steel. The floor inside the container is generally made of wood. Nearly all are manufactured in China.

The Container Revolution

The invention of the container in the middle of the twentieth century revolutionized maritime trade: by dramatically lowering transportation costs, container technology created new international markets for manufacturing goods and hence new trading partners. And yet the technology that underpins the container system is dead simple: it relies on a plain steel box, 20 or 40 feet in length, eight feet in width, and eight and a half feet in height. Because the box can be moved easily from train to truck—or from truck or train to ship—without ever disturbing the contents inside, its invention eliminated many of the labor costs associated with the repeated handling of maritime goods.

Howland Hook Reborn

Unbeknownst to most New Yorkers, the City of New York has not one but two container terminals. The first, the Red Hook Container Terminal in Brooklyn, sits opposite Governors Island and can commonly be seen from the Staten Island Ferry or other harbor craft. The second, New York Container Terminal, on the northwest shore of Staten Island, is all but invisible to the vast majority of city residents.

Then known as Howland Hook, the terminal was once the proud home of U.S. Lines, a company founded by Malcom McLean, the inventor of the container. But the company overbuilt and overspent, and after its bankruptcy in 1986 the terminal lay fallow for a decade. Reactivated by the Port

North Atlantic Port Comparisons
The New York/New Jersey region is by far the largest containerport in the North Atlantic, handling more than twice as many containers as Norfolk/Hampton Roads, its closest competitor.

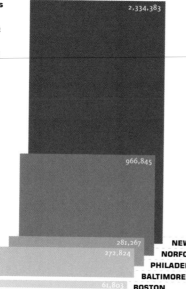

Loaded Twenty-Foot Equivalent Units (TEUs), 2002/2003

2,334,383 — NEW YORK/NEW JERSEY
966,845 — NORFOLK/HAMPTON ROADS
281,267 — PHILADELPHIA
272,824 — BALTIMORE
61,803 — BOSTON

The earliest containers were lifted between dock and ship by "ship's gear"—cranes installed permanently on the deck of the ship. While this avoided the expense of installing cranes at each port of call, as cargo loads grew new ways needed to be found to keep ships on an even keel during loading. The answer was the shoreside crane: set into railing along a berth, these cranes were designed to move along a ship's hull to collect or stow containers. As ships grew, so too did the cranes: modern cranes can lift containers up to 80 tons and reach across 20 rows of containers. And they are fast—the speediest container cranes can move between 50 and 60 containers an hour.

Container Cranes

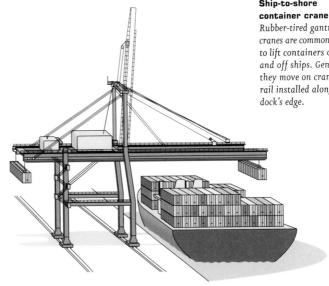

Ship-to-shore container crane *Rubber-tired gantry cranes are commonly used to lift containers on and off ships. Generally, they move on crane rail installed along a dock's edge.*

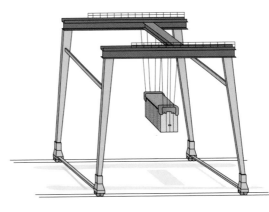

Rail-mounted gantry crane *Rail-mounted gantry cranes are used in high-density stacking operations. They maximize container terminal efficiency by allowing for close spacing of containers in the yard.*

Authority in 1996, it is now home to New York Container Terminal. Together the Port Authority and New York Container Terminal invested over $70 million to reopen the facility and are now investing another $350 million in new berths, cranes, and rail yards. The terminal is now the largest employer on Staten Island.

Ship's gear *Cargo in some places is lifted off the dock and put in the hold of the ship by the ship's own cranes.*

Even less visible than freight trains or cargo ships, air cargo is big business in the

Air Cargo

metropolitan region. Roughly 2.6 million tons of air cargo, worth over $140 billion, move through the three New York region airports each year, generating 85,000 jobs and creating $9 billion in economic activity. Together, JFK, Newark, and LaGuardia Airports account for roughly 25 percent of the nation's air imports and 17 percent of its exports. The overwhelming majority of this volume is international trade, much of it high-value or perishable goods poorly suited to the longer transit times associated with shipping.

The air cargo industry traces its origins back to the late 1940s, when the Port Authority took over the three regional airports: LaGuardia, Newark, and Idlewild (now JFK). Perhaps no event at the time was more formative for the industry than the Berlin Airlift which—from the summer of 1948 through May of 1949—moved hundreds of thousands of tons of food and millions of tons of coal to the besieged German city. The airlift proved to the world that the large-scale movement of freight by air was possible.

These early cargo movements were labor-intensive: each shipment was carried onto the plane and tied down using a series of canvas straps and nets. But just as it had done in the maritime industry, containerization would radically improve the economics of cargo travel by air: containers offered better space utilization inside aircraft and protected shipments from damage and theft. Along with the introduction of faster and larger jet aircraft, the new technology made moving freight by air both competitive and highly desirable for valuable and perishable commodities.

The rapid expansion in air freight that resulted from the introduction of the container put pressure on existing facilities at airports in New York and elsewhere. The aging warehouses that had initially handled the new form of cargo were inadequate to handle the expansion of the

Ever since construction for the city's largest airport began on the site of the Idlewild Airport, in 1942, freight transportation facilities there have continued to expand.

industry, and new air cargo facilities began to be constructed—generally in partnerships between a cargo carrier and the airport operator. Through the late 1960s and most of the 1970s, at JFK Airport and elsewhere, older on-airport cargo buildings shared by multiple tenants were razed to make room for terminals for individual cargo airlines.

In addition to the terminals themselves, new forms of container-handling equipment had to be devised. This equipment was designed to load cargo either through the front, or nose, of the plane or along doors near the back of the fuselage. Speed of loading and reduction of labor hours was key to a successful loading operation, regardless of whether the cargo was palletized or traveling in air containers.

For a number of years, New York—in particular JFK—was the premier gateway for air cargo into North America. However, the rise of the integrated carriers, such as Federal Express, led to dramatic changes in the industry. These carriers controlled the entire logistics chain, from the time an envelope left a customer's hands until it reached the recipient. And they sited their major sorting facilities in locations with relatively cheap labor and airport capacity, such as Memphis.

The rise of the integrated carriers affected the balance between the region's airports. With the construction of both FedEx and UPS facilities there, Newark now rivals JFK as a home to air cargo. In some places in the region the cargo business has grown beyond the airport; Springfield Gardens in Queens, for example, has proved a profitable base for a number of air cargo companies that have developed headquarters there.

Air Cargo Volumes *The rise of the express/overnight carriers has dramatically changed the fortunes of many of America's largest cargo airports, including those in New York. At one time, JFK was the nation's premier air cargo gateway, handling 50 percent of all goods flown in by air. Today its market share has been cut in half, and it ranks fifth in size behind Memphis, Anchorage, Los Angeles, and Miami.*

Air Cargo Volume (2004)

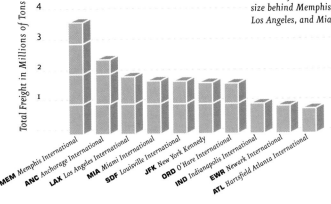

Air Cargo Historic Volume

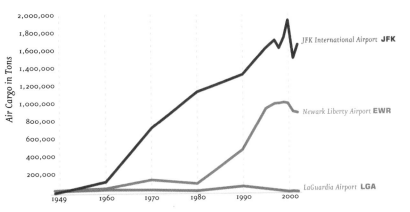

Air Cargo

Major Air Cargo Imports and Exports, 2004

(1000 metric tons)

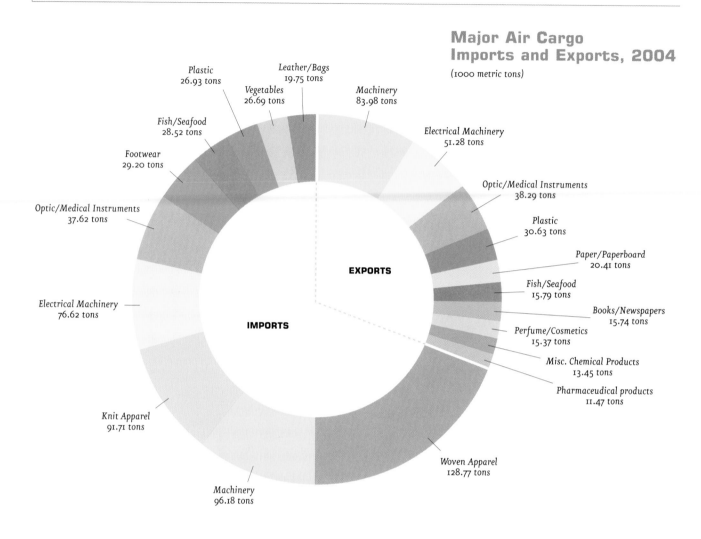

Plastic
26.93 tons

Leather/Bags
19.75 tons

Vegetables
26.69 tons

Machinery
83.98 tons

Fish/Seafood
28.52 tons

Footwear
29.20 tons

Electrical Machinery
51.28 tons

Optic/Medical Instruments
37.62 tons

Optic/Medical Instruments
38.29 tons

Plastic
30.63 tons

Paper/Paperboard
20.41 tons

Electrical Machinery
76.62 tons

EXPORTS

Fish/Seafood
15.79 tons

Books/Newspapers
15.74 tons

IMPORTS

Perfume/Cosmetics
15.37 tons

Misc. Chemical Products
13.45 tons

Pharmaceudical products
11.47 tons

Knit Apparel
91.71 tons

Woven Apparel
128.77 tons

Machinery
96.18 tons

Commodities

What sorts of things move by air? The simple answer is some of the most expensive and delicate cargos in the world—everything from Old Masters paintings to diamonds to racehorses. With cargos such as these, time is money, and a journey by air knocks weeks off the typical ship transit times. Equally important to the movement of these cargos is security: while pilferage and damage dropped sharply once the container was introduced to maritime trade, they still occur far more frequently on the water than in the air.

New York imports 83 percent of the country's diamonds, 55 percent of its arts and antiques, and 47 percent of its perfume and cosmetics. On the export side, the region exports two-thirds of all live lobsters sent out of the country and a full 40 percent of all seafood flown abroad by air. Unusual cargos that have recently made an appearance at the region's airports include baby chicks, helicopters, and spacecraft parts.

Air Cargo Facilities Today, the New York region's air cargo business is split largely between JFK and Newark, the latter being home to many of the largest "integrated carriers" such as Federal Express and UPS. (LaGuardia maintains a small share of the market, primarily catering to short- and medium-haul domestic services.) Some 1,000 cargo companies have a presence at one or more of the three airports, employing an estimated 15,000 cargo professionals.

Of the three, JFK is the largest and most diverse cargo center. Its Air Cargo Center is comprised of almost three dozen cargo-handling and cargo-service buildings as well as a U.S. Post Office facility. American, Lufthansa, Korean Air, Delta, Asiana and Polar Air Cargo are among the largest carriers at Kennedy—and facilities are expanding. Since 1992, roughly 1.3 million square feet of new warehouse space have been added at the airport.

An Overview of JFK

🔲 cargo building 🔲 other building

Northwest Airlines Cargo Northwest operates a highly automated facility geared to the needs of its primarily Pacific market.

DHL Danzas DHL Danzas is a major freight forwarder with a presence at the airport.

Nippon Cargo Air Lines Nippon Cargo Air Lines' new cargo terminal, at 175,000 square feet, can accommodate two 747 freighters side by side.

Halmar Cargo Center This building provides cargo storage and clearance services and incorporates all U.S. Customs JFK office operations.

Korean Airlines Korean Airlines, the largest Asian-based handler of air cargo, manages a new facility at JFK capable of handling 200,000 tons annually.

Japan Air Lines JAL has a 260,000-square-foot, $115-million cargo building near the airport's administrative offices.

JFK vs. Manhattan

JFK is by far the largest of the region's three airports. At just under 5,000 acres, it is roughly the size of Manhattan below Central Park and boasts 30 miles of roadway.

Air Cargo

Stowage Roughly half of the air cargo entering or leaving the New York region moves in the belly of passenger planes, much the way passengers' luggage does. The remainder is handled by dedicated cargo craft known as "freighters." With some exceptions, it generally moves on pallets or in air cargo containers, boxes designed with corners chopped off to suit the limited confines of an airplane's lower hold. A variety of container loading systems are in use to minimize the amount of manual labor involved in the loading operation.

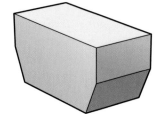

Specialized air cargo containers have been designed with rounded corners to take account of the shape of cargo holds on a variety of aircraft.

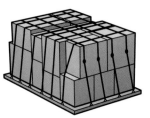

Some air cargo moves on pallets designed to fit either on the upper or lower deck of freighters or passenger planes.

Components of the Air Cargo System

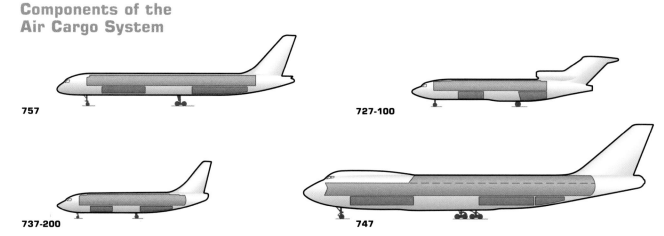

757

727-100

737-200

747

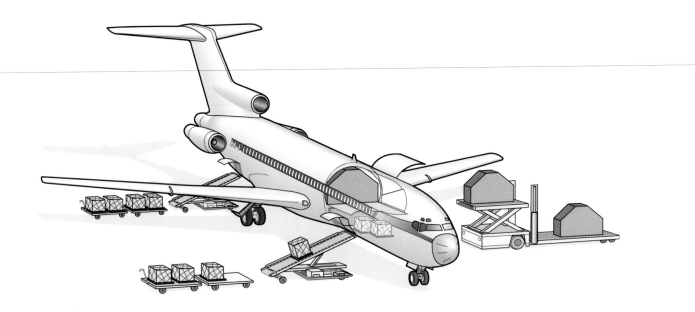

Express Carriers Over the past decade, the express package business has become a huge part of the air cargo market in New York and elsewhere in the country. Federal Express alone now accounts for a full 25 percent of all regional freight volume; UPS accounts for roughly 6 percent. In Newark, these companies—known as "integrated carriers"—absolutely dominate the air cargo market, accounting for 66 percent of the airport's cargo business.

Federal Express's Newark facility, for example, operates around-the-clock and handles roughly 400,000 packages each day. Hosting outbound flights in the morning and inbound ones in the evening, it is highly efficient; its staff can turn a plane around in an hour. A delayed departure is rare, thanks to a high-demand parts bank located near the airport that helps minimize time lost to mechanical failure.

FedEx on the Move

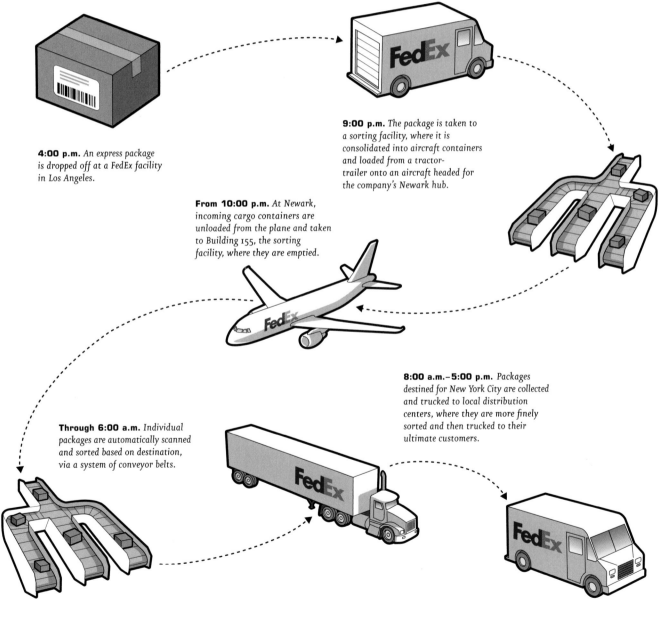

4:00 p.m. *An express package is dropped off at a FedEx facility in Los Angeles.*

9:00 p.m. *The package is taken to a sorting facility, where it is consolidated into aircraft containers and loaded from a tractor-trailer onto an aircraft headed for the company's Newark hub.*

From 10:00 p.m. *At Newark, incoming cargo containers are unloaded from the plane and taken to Building 155, the sorting facility, where they are emptied.*

Through 6:00 a.m. *Individual packages are automatically scanned and sorted based on destination, via a system of conveyor belts.*

8:00 a.m.–5:00 p.m. *Packages destined for New York City are collected and trucked to local distribution centers, where they are more finely sorted and then trucked to their ultimate customers.*

Since its founding as a trading post in 1624, New York has been a city of commerce and markets.

The earliest markets, in the seventeenth century, were comprised of pushcarts in lower Manhattan that sold vegetables, meats, and dairy products from local farms. In 1812, to reduce the congestion on downtown streets, the city's first real wholesale market—Washington Market—was created in an area bounded by Washington, Fulton, and Vesey streets. Washington Market prospered through the first half of the twentieth century, but ultimately congestion undermined its competitiveness as a wholesale destination, and the city closed it down in 1956.

Wholesale markets sprouted up in the outer boroughs as well. In Brooklyn, the Wallabout Market— the largest farmers' market in the borough—opened in 1884 adjacent to the Brooklyn Navy Yard. Following the site's acquisition for a dry-dock facility in 1941, the market relocated to Canarsie— where it continues to operate today as the Brooklyn Terminal Market.

The Bronx developed its own terminal market in the 1920s. Stretching along the Major Deegan highway between East 149th St. and the Macombs Dam Bridge near Yankee Stadium, the Bronx Terminal market was built as a receiving point for fruits and vegetables. It continues to operate today at a reduced level of activity than it once did, and may vacate its current site to make way for a large mixed-use real estate development.

The major produce markets have historically been supplemented by meat markets. In addition to one at Hunts Point, two smaller meat markets—the Brooklyn Wholesale Meat Market on the Brooklyn waterfront and the Gansevoort Meat Market in Greenwich Village—operate today. Relocated from Fort Greene, the Brooklyn Wholesale Meat Market today comprises over 150,000 square feet in two buildings.

Markets

The Gansevoort Market, as seen from Washington Street looking west, 1885.

Making a Fruit Salad

The New York region imports fruit from countries across the world. Bananas, many of them from Ecuador, are by far the largest fruit import.

Weight of trade in thousands of metric tons

🥥 Coconuts, brazil nuts, & cashew nuts, fresh or dried
🍇 Grapes, fresh or dried
🍍 Dates, figs, pineapples, avocados, etc., fresh or dried
🍊 Citrus fruit, fresh or dried
🍎 Apples, pears, and quinces, fresh
🍌 Bananas and plantains, fresh or dried

The Banana Pipeline

A bunch of bananas can be found at almost any corner deli in the five boroughs. In Manhattan delis, at the high end of the range, they sell for about 25 cents each; at sidewalk vendors and in the outer boroughs they can be found for considerably less. But it is very likely that those bananas—wherever they end up and however much it cost to buy them—started in the same place, came on the same ship, and moved over the same docks just days before.

Each year, some 100 million bananas are consumed in New York. The lion's share are grown in plantations in Central or South America and are moved in refrigerated ships to New York's port. One of the biggest points of entry is New York Container Terminal on the north shore of Staten Island, where the bananas—still very green—are unloaded for transport to wholesalers.

At the wholesalers' premises, many of them inside the Hunts Point Market, the bananas are moved to "ripening rooms" where they are kept at an average temperature of 58 degrees. From there they will be sold in different stages of ripeness to the retailers: a big retailer may want half green and half yellow bananas; a smaller retailer may want all yellow bananas for quick sale. The ripening process generally takes about five to eight days in total.

Markets

The Gansevoort Meat Market dates back to the turn of the last century, when 250 slaughterhouses and packing plants formed the core of the district. Today, about a dozen meat businesses remain—sitting not always comfortably beside galleries, clothing boutiques, and late-night clubs.

Each of these markets is small in comparison to the three best known and largest New York markets: the Hunts Point Produce Market and Meat Market and the Fulton Fish Market. Hunts Point is really more than a market: it is the country's largest wholesale food distribution center. Covering 329 acres, the complex comprises the Hunts Point produce market, a cooperative meat market, the broader "food distribution center," and a variety of private food distributors and wholesalers. Receiving daily deliveries by road, air, and rail, it serves an estimated 15 million customers in the tristate area—including 17,000 eateries and close to 3,400 street vendors in New York City alone.

The Fulton Fish Market was added to the roster of Hunts Point markets in 2005, when it moved to a new 430,000-square-foot building on the south side of the Bronx complex. The move from its historic location at the South Street Seaport, prompted by new federal regulations prohibiting the open-air sale of fish, is not expected to undermine its popularity as the clearinghouse for fish in New York: its 40 wholesalers currently sell about 220 million pounds (and 300 species) of fish each year, making it the world's second-largest central fish exchange (after Tokyo).

Metropolitan Markets

Bronx Terminal Market *Built in the 1920s as the city's first terminal market, the Bronx Terminal Market has more ethnic foods and more of a retail-like environment than Hunts Point.*

Gansevoort Meat Market *Although it houses only about a dozen businesses today, in its heyday a century ago the Gansevoort district had close to 250 slaughterhouses and packing plants.*

Brooklyn Meat Market *The Brooklyn Meat Market, located at 1st Ave. and 56th St., has been at this site since its relocation from Fort Greene in the 1970s.*

Brooklyn Terminal Market *Originally called Wallabout Market, the Brooklyn Terminal Market opened in 1884 right next to the Brooklyn Navy Yard. At the time, it was Brooklyn's largest wholesale farmer's market, closing only when the Navy Yard expanded after the start of World War II.*

A Night at the Fulton Fish Market

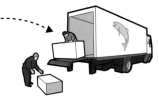

Between 9 p.m. and midnight, an estimated 500,000 pounds of fish in 80 or so trucks arrive at the market. Between 20 percent and 40 percent of this total has arrived by air via JFK.

After arrival at the market, registered unloaders take the fish off the trucks and organize it onto palettes. Larger loads can take up to an hour to unload completely.

The Geography of Hunts Point Market

Hunts Point Terminal Market
The Hunts Point Terminal Market, one of the premier produce markets in the world, is located on 126 acres within the Hunts Point complex. Its 55 fruit and vegetable wholesalers, organized as a coop, sell almost three billion pounds of fruit and vegetables a year, with revenues in excess of $1.5 billion. It consists of four large buildings—each about three blocks long.

Food distribution businesses In addition to the organized markets, there is a range of single warehouses for various food processing and distribution businesses.

Rail delivery Hunts Point receives regular rail service from agricultural producers via CSX Railroad. Its annual total of 3,000 railcars make it the single biggest user of rail service within New York City.

Fulton Fish Market The new Fulton Fish Market, at 300,000 square feet, is 35,000 square feet larger than its predecessor. The selling area is fully refrigerated, allowing displays of product without ice. Loading decks allow up to 20 trucks at a time to unload much more rapidly (two and a half hours) than was possible on the streets of Manhattan (five hours).

Hunts Point Cooperative Market The Hunts Point Cooperative Market, involved in the processing and sale of meat, is located on 60 acres and consists of six large refrigerator-freezer buildings. Some 47 independent wholesale food businesses make up the coop.

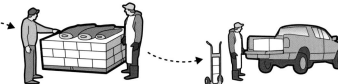

From 10 p.m. to roughly 1 a.m., the fish is delivered to the wholesalers' stalls. Some wholesalers will fillet the fish before putting it on display.

From around 3 a.m. to 6 a.m., the retailers arrive. Upward of 600 buyers will visit on a normal night, looking for the freshest fish and the best prices. The average customer spends about three hours at the site.

After the fish is bought, a licensed loader delivers the fish to the buyer's car or van. The busiest hours for the loaders are between 4 a.m. and 7 a.m.

POWER

Energy powers the city—its lights, its computers, and its telephones. Vertical New York is more dependent on it than most cities—for movement up and down in its buildings, as well as for mass transit, for heat in the winter, and for cooling in the summer. Together New Yorkers consume as much power each year as the entire country of Greece—and that total continues to grow.

Our energy infrastructure goes far beyond the power plants along the city's waterfront; without the gas pipelines feeding the city, both those plants and many industrial businesses would grind to a halt. The energy network stretches to the northernmost reaches of New York State; for while electricity is generated within the city, much is also imported over the state power grid. And it includes Manhattan's massive central steam system—the largest of its kind anywhere in the world.

New York City is known
to many as "the city
that never sleeps," where
the lights burn bright

In many ways, New York is where electricity was born. On September 4, 1882, two years after establishing a patent on his new electric lamp, Thomas Edison kicked off the operations of the Edison Electric Illuminating Company of New York at his new Pearl St. generating station. The station cast light on only a few buildings in lower Manhattan—its one generator produced power for 800 lightbulbs—but proved the viability and reliability of central power generation and distribution, and at a cost competitive to the gas lighting then in use. The *New York Times* reported that the light, which had recently been introduced into its offices, was "soft, mellow, grateful to the eye."

until the wee hours and the subway never stops. But it is more than just its reputation that New York's power infrastructure must maintain: an estimated five million air conditioners, seven million televisions, nine million cell phones, and two million personal computers are all dependent upon the uninterrupted supply of electricity to New York City.

Over a hundred years later, New York City's power grid is a marvel of modern engineering. High-voltage power delivered from generation plants is stepped down by some 33,000 transformers and distributed through the world's largest underground electric cable system, featuring over 80,000 miles of underground cable—enough to encircle the globe three and a half times. Access to the cables is provided by 250,000 manholes across the city.

Electricity

New York's is not just one of the largest systems of delivering electricity in the world—it is one of the most reliable. The city's power delivery infrastructure is 10 times as reliable as the average U.S. system, thanks to a reliable design with backup systems and multiple feeders to individual neighborhoods. Power failures are rare within the city; when they do occur, they are generally prompted by events caused hundreds or thousands of miles away.

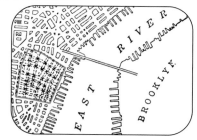

In 1882, the Edison Electric Illuminating Company of New York, predecessor of today's Con Ed, lit up one square mile in the Pearl St. district downtown.

The first successful generator was Edison's "Jumbo No. 1." It weighed 27 tons and was built in 1881 at the Edison Machine Works in New York City.

Consumption The city consumes vast quantities of electricity each day, regardless of the season. Its summer peak load is 11,000 megawatts (one megawatt represents the amount needed to power 1,000 homes). That puts it on a par, as an energy consumer, with many countries around the world—using slightly more than Chile and Portugal and slightly less than Switzerland and Austria.

But while New York as a whole consumes a lot of electricity, New Yorkers as individuals are not huge or wasteful consumers of energy. To the contrary: the average New Yorker consumes about 2,000 kilowatt hours per year—compared to 4,000 for the average American. Even within Con Edison's service territory, the difference is notable: while 450 kilowatts are used each month in the typical Westchester home, the average New York City customer's home uses only 300.

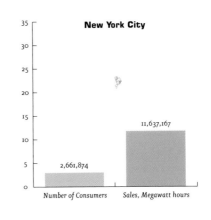

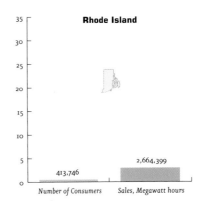

Comparing Power Consumption

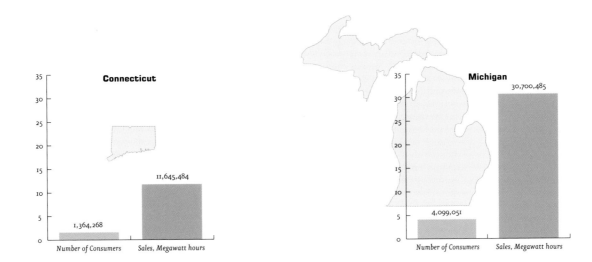

Electricity

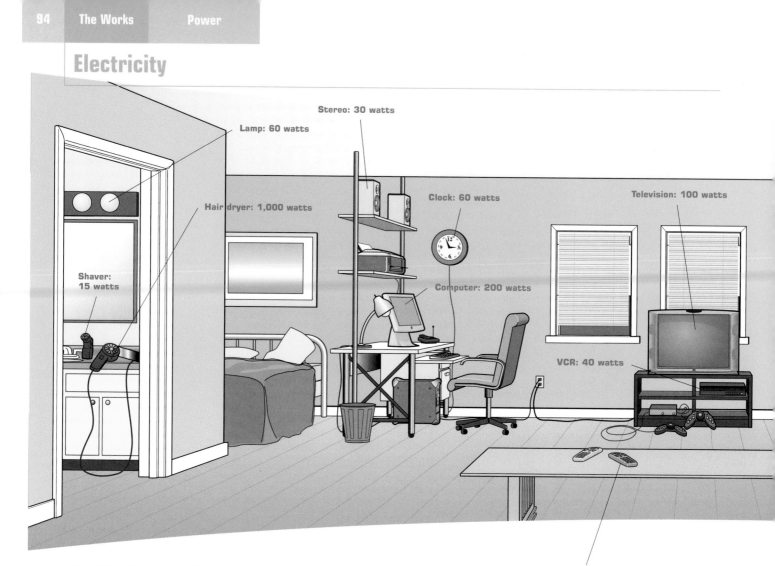

Lamp: 60 watts

Stereo: 30 watts

Clock: 60 watts

Television: 100 watts

Hair dryer: 1,000 watts

Shaver: 15 watts

Computer: 200 watts

VCR: 40 watts

New York City's power demands are not just big—they are growing. Over the last five years, peak demand within the city has grown at an annual rate of between 1 and 2 percent; this represents an increase of over 800 megawatts. And the trend shows no sign of abating; recent studies suggest that by 2008 the city will require an additional 2,600 megawatts of electricity capacity locally to meet forecast demand.

The city's appetite for power is in part a function of its heavy reliance on the subway and on elevators, cornerstones of its distinctive urban lifestyle. But it is also a function of its insatiable appetite for home appliances and electronic forms of communication. Air conditioning, for example, has grown dramatically in the last decade; it is now estimated that 61 percent of city homes are air-conditioned, up from 56 percent eight years ago. The use of computers, cell phones, and televisions has similarly grown, as costs have dropped and technology has improved. And while it is true that individual appliances may be getting more energy efficient, their rapid proliferation outweighs any energy savings that better manufacturing has achieved.

Remote Power *One of the least visible elements in the growing demand for power in New York City is the use of "remotes," or remote-controlled devices. Appliances that are turned off using a remote control, as well as those that are in "sleep" mode (such as many computers), are always on—whether they appear so or not. A remote control doesn't turn an appliance off and on—it just activates an "instant-on" device that always remains on. So for every remote activated, there's an electrical appliance drawing electricity from the grid in "sleep" mode.*

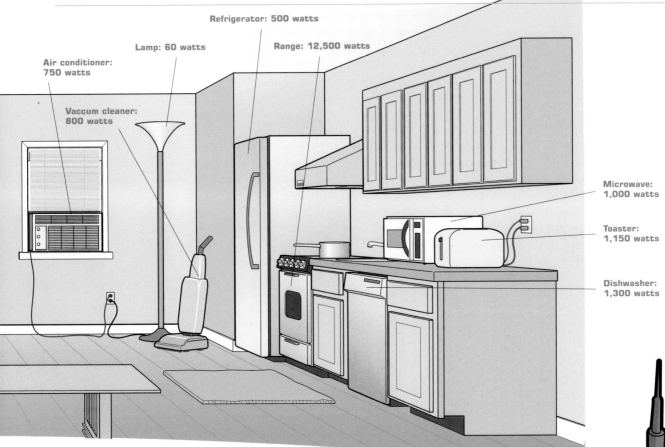

Air conditioner:
750 watts

Vaccum cleaner:
800 watts

Lamp: 60 watts

Refrigerator: 500 watts

Range: 12,500 watts

Microwave:
1,000 watts

Toaster:
1,150 watts

Dishwasher:
1,300 watts

Annual Electricity Use by City Offices and Public Buildings
(in Megawatt Hours)

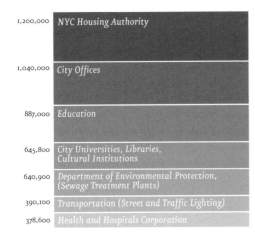

1,200,000	NYC Housing Authority
1,040,000	City Offices
887,000	Education
645,800	City Universities, Libraries, Cultural Institutions
640,900	Department of Environmental Protection, (Sewage Treatment Plants)
390,100	Transportation (Street and Traffic Lighting)
378,600	Health and Hospitals Corporation

Public Power New York City itself is a big consumer of electricity, accounting for roughly one-tenth of all the electric power used within the city. Nearly all city power is supplied by the New York Power Authority (NYPA), a state-owned public authority which also sells power to the MTA and Port Authority.

Lighting the Empire State Building The top 30 floors of the Empire State Building were first illuminated with floodlights to mark the beginning of the New York World's Fair in 1964. Colored lighting was first introduced 12 years later, to celebrate the bicentennial. Today, over a thousand fluorescent bulbs in five different colors (yellow, white, red, green, and blue) on the building's topmost mooring mast can be changed at the flick of a switch. Inside the building, approximately 2.5 million feet of electrical wire carry some 40 million kilowatt hours to the building's tenants each year.

Electricity

Generation At its simplest, electricity production relies on the rotation of a turbine—a large shaft with blades—at a power plant. What forces the turbine to turn can vary; gas combustion and steam are the most common, but wind or water can work as well. The turbine is connected to a long shaft that turns giant electromagnets, which are surrounded by coils of copper wire. The rotation of these magnets creates a magnetic field, which generates electric current in the wires.

To produce electricity, a steam or other form of turbine will rotate a shaft connected to a generator.

Coils of copper wire attached to the shaft are rotated within giant electromagnets to create a magnetic field, causing electrons to flow and produce electricity.

Electricity moves from the turbine generator to the transmission lines.

Inside a Power Plant

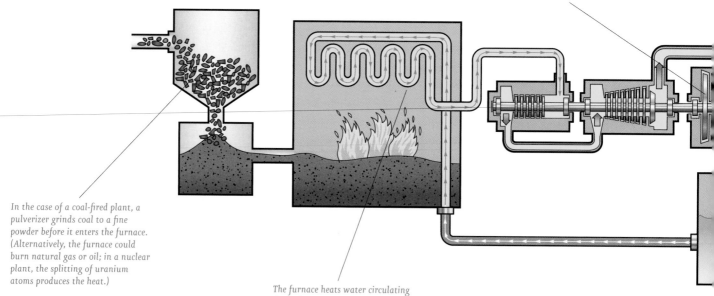

High-pressure steam spins the turbine. As the steam loses pressure, it flows past different sizes of blades, each angled to extract the maximum energy from the steam.

In the case of a coal-fired plant, a pulverizer grinds coal to a fine powder before it enters the furnace. (Alternatively, the furnace could burn natural gas or oil; in a nuclear plant, the splitting of uranium atoms produces the heat.)

The furnace heats water circulating through pipes to make steam.

Defining Energy Even a basic understanding of electricity requires familiarity with a variety of terms. "Current" refers to the flow of electrons through a wire, while "voltage" refers to the force behind the current, or how hard the electrons are being pushed through the wire. Most electronic devices, and most of the nationwide grid, run on alternating current, which means that the current and voltage move in back and forth pulses rather than a steady flow (direct current). The combination of current and voltage is represented by "watts," a basic measure of how much work the electricity can do.

The central shaft of the turbine is connected to a generator that turns the mechanical energy of the spinning shaft into electrical energy.

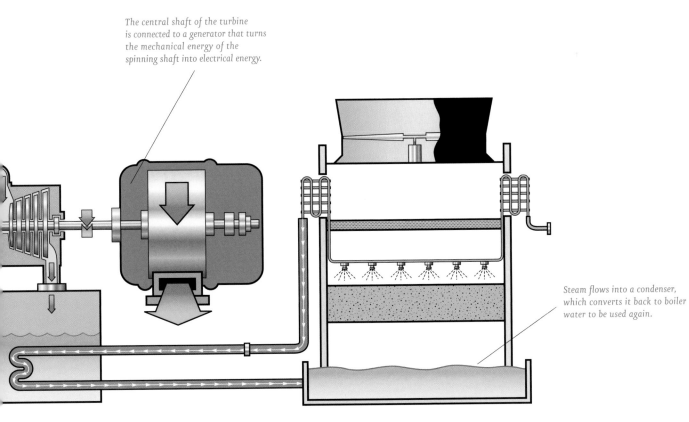

Steam flows into a condenser, which converts it back to boiler water to be used again.

Electricity

Power Plant Locations

Indian Point *The out-of-city plant relied on most heavily is Indian Point, a nuclear plant run by the Entergy Corporation and located 35 miles north of Manhattan on the Hudson River in Westchester County. Three pressurized-water reactors make up the facility, although only two—Indian Point 2 and 3—are in operation. Together they produce about 2,000 megawatts of electricity —roughly 5 percent of New York State's power, some of which ends up in New York City.*

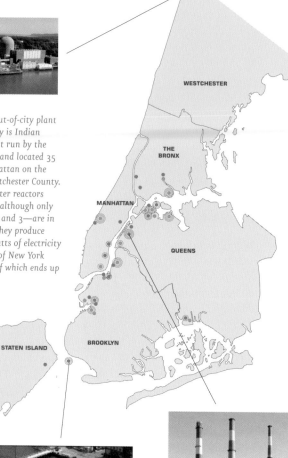

Legend:
- ■ Existing
- ■ Proposed
- • < 100 MW
- ◉ < 500 MW
- ◉ < 1000 MW
- ◉ < 5000 MW

Power barges, *which normally function as emergency or temporary power sources, have become something more permanent in New York City. Located on parts of the Brooklyn waterfront for 30 years, they are primarily used to produce power to meet peak demand. The Gowanus site—with 32 combustion units—is the world's largest floating power plant.*

"Big Allis" *remains the star attraction at New York's largest power plant, Ravenswood, in Long Island City. Otherwise known as "Generator #3," Big Allis was the world's largest electricity generating unit when placed into service by Con Ed in 1965. Nicknamed for the manufacturer of the turbine, the Allis Chalmers Corporation, she can produce one million kilowatts (100 megawatts) of electricity.*

Power Plants Most of New York City's power is produced at central generation plants within the five boroughs. Four are located in Queens and account for roughly half of the energy output in the city; the remainder are spread across Staten Island, the Bronx, and Brooklyn. At one time, all of the generation system was owned and operated by Con Edison. In 1999, however, as part of the restructuring of the electricity industry, Con Ed retreated from the electricity-generation business and the plants were sold to private companies; NRG Energy, KeySpan, and US Power Generating Company own them today.

The central generation plants are supplemented by smaller generators to meet peak demand. Since 1975, the city has relied on a fleet of power barges sited primarily in the Gowanus Bay and at Sunset Park along the Brooklyn waterfront. These function as mini natural-gas–powered generating plants. The fleet was expanded to six additional sites during the summer of 2001, in response to warnings that the metropolitan area could face power shortages.

Not all of the city's power is produced within the five boroughs. Indian Point, a nuclear plant in Westchester County, produces up to 20 percent of city demand. In addition, the city draws power produced in plants located farther upstate from the New York State grid as well as power produced by plants in New Jersey, Connecticut, and Pennsylvania.

At present, because of the design constraints related to the city distribution grid, limits are placed on how much power may be provided from plants outside of the city. The city must have within its boundaries capacity to produce 80 percent of the power it is forecast to need at any given time; the remaining 20 percent may be delivered through the transmission system.

The State Grid New York State's power grid is among the most complex and congested in the nation, involving more than 330 generating plants. Coordinating the flow of electricity from this diverse group of producers requires a round-the-clock effort by hundreds of people employed by an organization known as the New York Independent System Operator, or ISO.

One of the primary jobs of the ISO is to maintain a competitive wholesale market for electricity. To balance supply and demand, the ISO runs what is known as a "last-price auction," matching sellers and buyers of electricity on a day-to-day basis across the state. Each morning, it accepts offers from generators that expect to have energy to sell the following day. At the same time, it takes bids from utilities looking to buy energy to meet the following day's demand. After sorting the offers by price, NYISO selects supply offers until all purchase bids are met; the last supply offer sets the price that all buyers pay and all sellers receive.

It is a never-ending job. Since electricity cannot be stored, the amount of energy coming into the system and the amount moving out must always be in sync or it will negatively impact the "power quality," which can affect the operation of electrical equipment. To ensure the system stays in balance, the ISO holds conference calls throughout the day with generators to monitor supply and matches them with requests from Con Ed and other wholesale buyers. Last-minute adjustments are made roughly every six seconds.

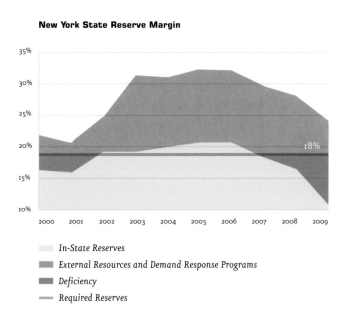

New York State Reserve Margin

- In-State Reserves
- External Resources and Demand Response Programs
- Deficiency
- Required Reserves

While most of New York State's power needs are met from in-state generation, a certain amount must be imported from other places to meet the 18% reserve margin requirement.

Managing the Peaks If a heat wave looks to be pushing demand for electricity unexpectedly high, the ISO will issue a "max gen pickup alert," which mandates each of 10 active power plants and hundreds of smaller generators within the city's grid to run at maximum output. It will simultaneously try to increase the amount of energy flowing into the city from upstate and from New Jersey through the transmission lines serving the load pocket. If these two actions don't appear to meet demand, emergency procedures are put in place: participating office buildings will be asked to reduce light, elevator usage, and air conditioning, and a 5 percent voltage reduction—a limited and barely noticeable brownout—will be implemented. Finally, if demand continues to exceed supply, the ISO will order rolling blackouts and Con Ed will be requested to reduce demand to preserve the integrity of the network as a whole.

Electricity

Alternative Energy Sources

A number of forms of alternative energy are available to supplement traditional fossil-fuel sources provided to the city. For example, hydroelectric power provides roughly 4 percent of the energy consumed by New Yorkers, and wind power just under 2 percent. For New York State as a whole, these alternative sources of energy are much more significant: an estimated 18 percent of New York State's power needs comes from hydroelectric power. Some of this power is produced in Canada; the balance is produced in upstate New York.

Some of the most interesting experiments with alternative energy sources in New York involve fuel cells. Like batteries, fuel cells use an electrochemical process (rather than combustion) to convert chemical energy into electricity and hot water. Typically, the chemical energy comes from the hydrogen in natural gas supplied as fuel. Because this gas is not burned, there are no significant emissions.

Although fuel cells are more expensive than conventional means of producing energy—by roughly two or three times—they have already begun to appear in the metropolitan area. The New York Police Department's Central Park precinct, for example, relies on a 200-kilowatt fuel cell outside its building for all of its electricity. Although its cost (about $900,000, or $4,500 per kilowatt) is higher than a diesel generator of similar capacity ($800–$1,500 per kilowatt), it is considerably less polluting and provides independence from any interruptions to the New York State power grid.

A Sampling of Green Power

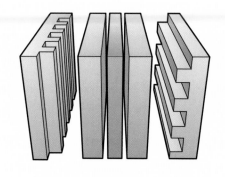

Biomass *Biomass refers to producing electricity from decomposing or burning organic matter, including gas from landfills.*

How to Buy Green

There are several ways to buy green power in New York. As a building owner, it is possible to contact a company to install solar panels on the roof of the building. As a resident, it is somewhat easier. Two companies sell wind power online: Community Energy and Madison Wind Project. An individual consumer can buy a block of kilowatt hours on line ($2.50 for 100-kilowatt hours). The power is generated at a wind farm and placed in the power pool used by Con Ed to distribute power to New York City consumers. While the specific energy purchased cannot be delivered to any particular home, the purchase decreases the amount of traditional, fossil-fuel–generated power in the overall state mix.

Fuel Cells A fuel cell has three basic components. The first is the fuel processor, which combines gas with steam generated by the system to produce a hydrogen-rich fuel mixture. As this fuel mixture moves into the next component of the cell, the power section, hydrogen atoms in the fuel mixture split into protons, which combine with the oxygen to form water, and electrons, which flow through a separate electrical circuit to create direct current. Within the third cell component, the power conditioner, the direct current power produced is converted to alternating current for use by electrical devices.

Wind *Electricity is produced by windmills, which emit no pollution and do not require large expanses of land.*

Geothermal *Geothermal energy involves converting hot steam or water below the earth's surface into electricity in volcanic areas, or using the earth's natural temperature to cool or heat water for air conditioning, heating, or other purposes.*

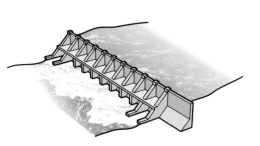

Hydroelectric *Water power can also be used to generate electricity. In most cases, turbines are placed at strategic points to capture the kinetic energy of moving water.*

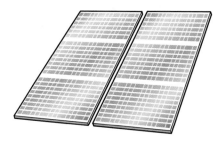

Solar *The sun can be used to generate electricity in two ways: photovoltaic and thermal. Photovoltaic panels use a semiconductor material to turn sunlight directly into electricity. With solar thermal technology, the sun's energy is most often used to produce hot water.*

Electricity

Distribution

Early power plants were small and localized, generating power primarily for immediate surrounding areas. As demand for electricity spread, however, it became clear that large generating plants could more efficiently meet the new levels of demand. For these centralized facilities to meet the needs of a broader geographic base of users, transmission lines were constructed to carry power from plants to residential, industrial, and commercial customers. Eventually, to increase efficiency and reliability even further, these transmission systems were linked with one another into a nationwide bulk power grid.

Today, the nation's power is delivered over an efficient network of high-voltage transmission lines connecting hundreds of generation locations and load centers. There are three major interconnected grids: the Eastern Interconnection, covering the East Coast, the Plains states, and Canada's maritime provinces; the Western Systems Coordinating Council Interconnection, which includes the West Coast, mountain states, and western provinces; and the Electric Reliability Council of Texas, which covers Texas alone. Each grid is composed of transmission lines operated by a variety of owners, from regulated utilities to private companies to federal power corporations. The grid aims to improve reliability and lower costs by offering its participants the opportunity to both buy and sell power with one another as well as use alternative power paths in an emergency.

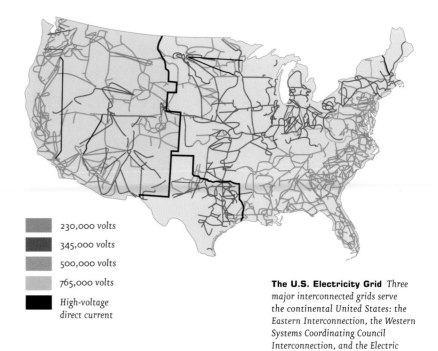

■	230,000 volts
■	345,000 volts
■	500,000 volts
■	765,000 volts
■	High-voltage direct current

The U.S. Electricity Grid *Three major interconnected grids serve the continental United States: the Eastern Interconnection, the Western Systems Coordinating Council Interconnection, and the Electric Reliability Council of Texas.*

Power Distribution

1. *Electricity is sent out from the power plant where it is generated.*

7–25 kilovolts

60–500 kilovolts

200 kilovolts

2. *It moves immediately to a transmission substation at the power plant, where large transformers increase the voltage from that produced by the generator (anywhere from 2,300 to 22,000 volts) to that needed to travel a long distance to end users (generally between 230,000 and 345,000 volts, but occasionally as high as 765,000 volts).*

The Northeast Energy Market **In New York State, the area controlled by the New York Independent System Operator has a peak load of over 30,000 megawatts and is interconnected with the energy markets of New England, PJM (Pennsylvania, New Jersey, and Maryland), the Independent Electricity Market Operator of Ontario, and Hydro-Québec. Transmission lines allow a certain amount of power to move back and forth between markets at any time during the day or night.**

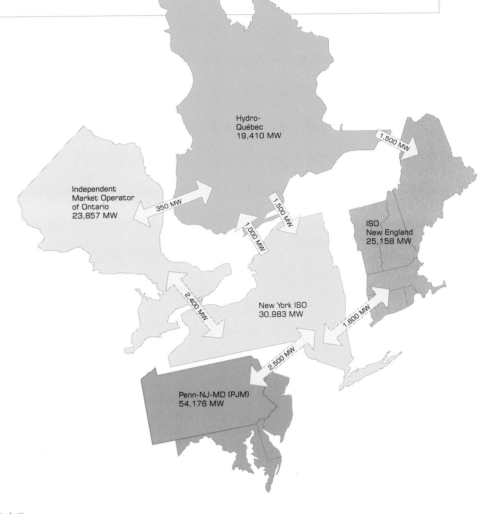

Hydro-Québec
19,410 MW

1,500 MW

Independent Market Operator of Ontario
23,857 MW

350 MW

1,500 MW

1,000 MW

ISO New England
25,158 MW

New York ISO
30,983 MW

2,400 MW

1,600 MW

2,500 MW

Penn-NJ-MD (PJM)
54,176 MW

3. *High-voltage transmission lines, normally characterized by tall steel towers, carry the power toward the city—typically up to 300 miles.*

4. *For New York City, power enters the city at 345,000 volts, 138,000 volts, or 69,000 volts. As it approaches its destination, it is "stepped down" at an area substation to a lower voltage (typically 13,000 volts) for movement onto the distribution grid, which is largely underground.*

5. *Power is sent out from the substation on a primary feeder, destined for a street transformer (mounted on poles or underground), which will step the power down again to the 120-volt or 220-volt level required for delivery to homes or businesses.*

12–35 kilovolts

120–240 volts

6. *Power leaving the street transformer travels underground on secondary feeders or a secondary main, in and out of manholes and often through a service box, before heading for a specific building.*

7. *Once inside a building, electricity generally goes to a service panel, which controls the building's power distribution to individual apartments.*

Electricity

Substations and Transformers Because power must travel at high voltages to cover the distance between generating plant and end user, yet be delivered at lower voltages for consumption, a system of stepping up and stepping down voltage is relied upon in nearly all major distribution systems. In general, voltage levels are determined by how far the electricity must travel and how much is desired. Substations located adjacent to the power plant and en route to the consumer are responsible for boosting or reducing power levels accordingly.

Most electricity headed for New York is generated at roughly 20,000 volts. In a fashion similar to the way that a pump can build up the pressure of water in a hose, it gets transformed up to a much higher voltage (69,000 to 765,000 volts) at a transmission substation adjacent to the generating plant; this allows it to be carried efficiently over long distances. At this higher voltage, the electricity is sent out along long-distance transmission lines, which may include lattice towers, steel poles, or simple double wooden poles.

As electricity approaches the area where it will be used, it enters an area substation, where it is transformed down again, to 13,000 volts, and sent out on primary feeders in the city distribution system. Substation transformers generally consist of a core and coils immersed in oil in a steel tank (the oil serves to both insulate the core and keep it reasonably cool for operations). In some cases fans or pumps, and on occasion even water spray systems, might be employed as additional measures to ensure the dissipation of heat.

Ultimately, primary feeders emerging from the area substation bring power to a street transformer, where it is converted down to domestic or commercial voltages (120 or 220 volts) and sent out on a secondary feeder to its destination.

At the Mercy of Marcy For almost half a century, the New York Power Authority's Marcy substation in upstate New York has transformed hydropower coming in on a line from Canada at 765,000 volts to power going out at 345,000 volts along two overhead transmission lines headed south.

One transmission line runs from Québec to New York City, passing directly through Albany. This line, generally referred to as Marcy South, is the state's most congested high-power transmission line.

A switch at Marcy controls the flow of electricity onto the Catskill and Albany lines. A pair of devices called "statcom" reshape the alternating current in each line, making it possible to pump power from one line to the other.

A second line, which generally has excess capacity, winds its way to New York City via the Catskills.

At the Marcy substation, a transformer drops the 765,000-volt flow coming in from Québec to a less sizzling 345,000 volts.

Transforming Power Area substations receive high-voltage power from long-distance transmission lines and step it down, generally to 13,000 volts, to be sent out on primary feeders throughout the city.

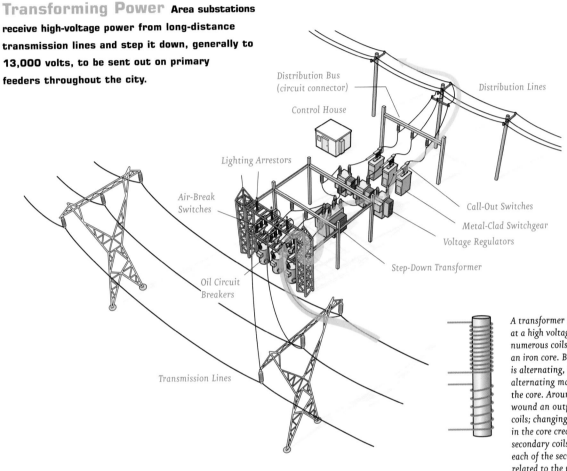

Distribution Bus (circuit connector)

Control House

Lighting Arrestors

Air-Break Switches

Oil Circuit Breakers

Transmission Lines

Distribution Lines

Call-Out Switches

Metal-Clad Switchgear

Voltage Regulators

Step-Down Transformer

A transformer takes in electricity at a high voltage and sends it through numerous coils wound around an iron core. Because this current is alternating, it provokes an alternating magnetism, or flux, in the core. Around this core is also wound an output wire with fewer coils; changing the magnetism in the core creates a current in these secondary coils. The voltage in each of the secondary coils is directly related to the primary voltage by the number of turns in the primary coil divided by the number in the secondary.

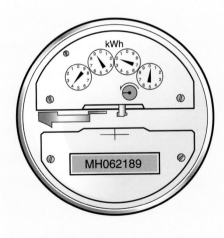

Metering Electricity Electricity meters have traditionally consisted of a number of circular dials, generally either four or five of them lined up in a row. The hands on a given dial often turn in the opposite direction to the dial next to it, either clockwise or counterclockwise. In general, to identify how much electricity is used, it is necessary to subtract only the earlier reading on the meter from the current reading. An increasing number of meters are now digital.

Electricity

To carry electricity, copper cables are often insulated in layers of lead and rubber. Larger pipes, up to a foot wide, feature cables bundled together and surrounded in oil under pressure to minimize movement and damage to the wires as a result of changes in power levels.

Wire Mania *Although Thomas Edison's first experiment with electricity distribution relied on subterranean tunnels, nearly all other communications and power infrastructure was located aboveground in the late nineteenth century. Most commonly, a 90-foot-tall pole could hold up to 24 cross ties, each of which could carry up to 20 wires. Although the state legislature ordered wires be placed underground as early as 1884, it was only after the Blizzard of 1888 left the city a tangled mess that the regulation was enforced and utility wiring was moved underground.*

Wiring the Street Unlike most parts of the country, New York City's local power distribution network is largely underground. Most of the transformers used to step down the voltage to the level required by residential customers are located in underground rooms or vaults, rather than on pole tops. Running cable from the substation under the street to the vault, and from there to individual buildings, would be impossible without the invention of the manhole.

At their simplest, manholes represent the start and end points of a run of cable—often 200 feet—and the place where the end of one cable is joined to the start of a second. In general, they measure about 12' x 8' or smaller, with a depth of about eight feet. Despite being covered most of the time, they are far from watertight; quite often they are filled with watery muck between visits from technicians. To clean them out, unless the mud is toxic, a "flush truck" will regularly flush water through the manhole and then vacuum it out—leaving the hole ready for service.

A variety of electrical wires flow through city manholes and service boxes. They range in size from 1/4 inch to three inches in diameter, carrying voltages anywhere from 120 volts to 13,000 volts. The high-voltage cables are the widest, and they feature oil inside a lead pipe surrounding insulated copper wires. The oil is kept under pressure to prevent the insulation around the wires from becoming loose as a result of the swelling and shrinking associated with varying levels of power throughput. These high-voltage cables, should they need repair, can take up to 12 hours to splice together.

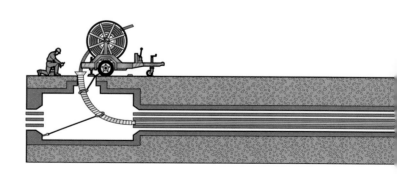

Inside a Manhole

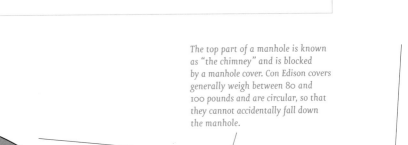

The top part of a manhole is known as "the chimney" and is blocked by a manhole cover. Con Edison covers generally weigh between 80 and 100 pounds and are circular, so that they cannot accidentally fall down the manhole.

The walls of the manhole are lined with cables on braces, which jut out from the walls. The braces support what are known as "crabs," units that accommodate several cables coming in and others going out.

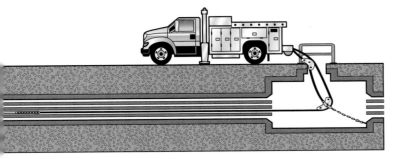

Laying cable generally involves a steel line being pushed from one manhole, over the length of a duct, to a second manhole. There, the end of the line is attached to a steel rope and winch, located in a specially designed truck parked above the hole. The electrical cable is then attached to the steel rope and pulled through.

Electricity

Blackouts New York City's electricity infrastructure is generally considered among the most reliable in the country. In particular, studies have suggested that Con Ed's reliability is four to nine times better than the national average. This perhaps should not be surprising, given how critical system reliability is to mass transit and apartment life and how protected most of the distribution grid is from the weather.

But when the city's electricity grid fails, it does so in spectacular fashion. People are trapped in subways, busy intersections are jammed with traffic, thousands of people must be rescued from elevators—urban life as we know it essentially comes to a halt. Over the last fifty years, complete failure has happened three times: in 1965, 1977, and 2003.

Each of these blackouts was caused by problems outside of the city. In 1965, a transmission line relay failed in Ontario, forcing generating companies to shut down plants to protect their equipment. In 1977, a thunderstorm took down lines north of the city, leaving the city and Westchester islands of darkness in an otherwise well-lit metropolitan region. Most recently, in August 2003, power failures as far away as Ohio triggered a series of shutdowns that affected much of the eastern seaboard and Canada.

In each case, processes were established to ensure against similar failure in the future—yet each blackout was more spectacular than its predecessor.

In general, however, blackouts in New York are rare. Con Edison's distribution system is designed so that if a local substation or generator fails, the rest of the grid will immediately supply power to prevent a deterioration in service; as a result, localized problems rarely spread. Likewise, brownouts—in which operators reduce voltage to feeder lines to avoid overload—mean little to the average residential customer, other than perhaps a slightly slower elevator.

The 2003 Northeast Blackout Timeline
Although the series of failures in Ohio and Indiana that were the source of the 2003 blackout occurred over a period of hours, the shutdowns of power plants they triggered across the Northeast occurred within seconds of each other.

12:00 p.m.   1:00 p.m.

12:08 p.m.: A transmission line in Indiana trips. Four minutes later, a second power line fails.

1:31 p.m.: A generating unit at a power plant in northern Ohio, owned by First Energy, fails.

Returning to Power

1. A power plant must be brought on line slowly, first producing power in small increments. Starting fossil fuel and nuclear plants can take time, as considerable energy is needed to turn enough water to steam in order to begin generating power.

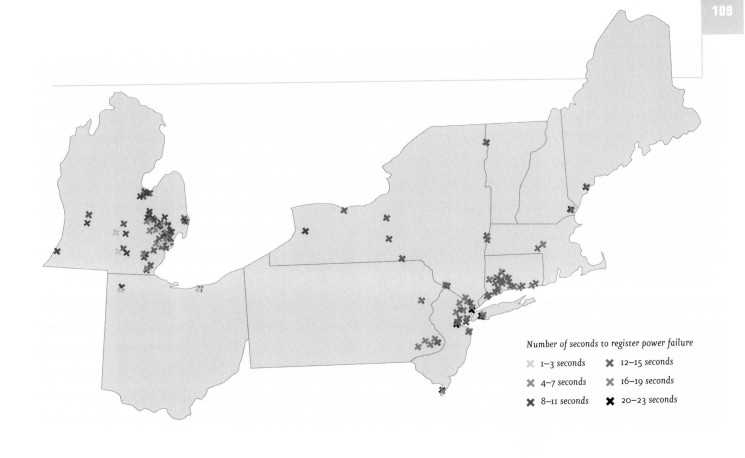

Number of seconds to register power failure

- ✖ 1–3 seconds   ✖ 12–15 seconds
- ✖ 4–7 seconds   ✖ 16–19 seconds
- ✖ 8–11 seconds ✖ 20–23 seconds

2:00 p.m.   3:00 p.m.   4:00 p.m.   5:00 p.m.

3:05 p.m.: First Energy loses one line in northeastern Ohio. A second one follows 25 minutes later.

3:30 p.m.: Utilities in the Cleveland area notice a severe drop in voltage.

3:30–4:00 p.m.: Increasing loads due to earlier failures cause a series of lines to fail in northeastern Ohio. Power begins to flow from Michigan to northern Ohio to make up the gap.

4:09 p.m.: More lines fail in Ohio, and the state begins drawing 10 times its normal amount of power from the Michigan grid.

4:10 p.m.: The load being pulled down by Ohio causes 30 transmission lines in Michigan to fail. Power from Canada is now being drawn into Ohio.

4:11 p.m.: The surge caused by the Ohio and Michigan failures spreads throughout the Northeast, causing power facilities across New York and points south to automatically shut down.

2. As small amounts of electricity are produced, limited numbers of customers may be brought on line by activating certain feeder lines.

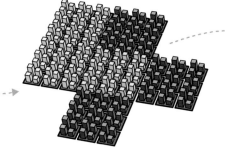

3. Electricity demand is added, neighborhood by neighborhood, until roughly half of the generator's capacity is being produced.

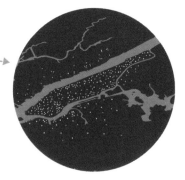

4. At that point, the plant can be connected first to other plants in the area producing at a similar level and then to the wider regional and state grids.

The use of gas as a form of power in New York City dates back some two hundred years, though

in the last fifty its sources and its delivery channels have changed dramatically. Originally produced by burning coal, the commercial provision of gas was big business in the five boroughs throughout the nineteenth century. The New York Gas Company, Con Edison's predecessor, was founded in 1823, nearly 60 years before

Natural Gas

Thomas Edison's Pearl St. experiment with electricity. By 1900, however, roughly 30 electricity-generating and -distributing companies were competing to provide light to Westchester County and New York City and gas—as a lighting source—was on its way out.

One hundred years later, commercial gas production has all but disappeared from New York. Instead, the city, its residents, and businesses rely today almost exclusively on natural gas—the cleanest-burning fossil fuel, and among the most plentiful. Thanks to the construction of thousands of miles of gas pipeline across the United States in the 1950s and 1960s, gas can now cost-effectively be brought to the East Coast from the Gulf of Mexico and from remote parts of Canada. Its popularity has soared, both as a fuel for use in the home and as a relatively clean fuel for generating power at plants across the region.

The term "natural gas" refers to hydrocarbon gases, primarily methane, which are generally found deep below the earth's surface. In most cases, it has been trapped there for hundreds of thousands of years, the product of decaying plant and animal matter. Often found alongside oil reserves, it is produced by drilling deep into the earth's crust and is brought up to the

A Century of Gaslight The first commercial uses of gas in New York relied exclusively on gas produced from burning coal. New York's first gas plant was built in lower Manhattan in 1823, and within five years the city had begun using gas to light its streets. By midcentury, gas lighting had spread to residential use and it would remain popular until the turn of the last century—when electricity became the fuel of choice and streetlights began to be converted to electricity.

surface by a well or pump. Once processed, with impurities and moisture removed, it is sent out through a series of pipelines to end users. Along the way, it can be stored in underground caverns; alternatively, it can be "frozen" into a liquid for easier transport and storage.

New York State is a big user of natural gas, accounting for roughly 6 percent of the country's consumption. Within the city itself, there are four primary groups of gas users: residential users (stoves, water heaters, and clothes driers); commercial users (restaurants, hotels, and hospitals); industrial users (heating processes, steam generation); and electric utilities (power generation). Factories and electric power plants may purchase their gas directly from the pipeline via a supplier, while remaining customers generally buy from a local distribution company. Two companies share delivery responsibility within the city: Con Ed serves Manhattan, the Bronx, and parts of Queens, while KeySpan serves Staten Island, Brooklyn, and the remaining parts of Queens.

Entering the City *Most gas delivered to New York City and Long Island travels through one of four interstate pipelines—the Iroquois, the Tennessee, Texas Eastern (and its Algonquin Pipeline affiliate), or Transco.*

□ *Receipt Points*
▪ *Bidirectional Gate Station*
■ *Pipeline Interconnections*

Gas Pathways *Gas is delivered to New York from the Gulf and western Canada by a series of pipelines of varying capacity. Over the last several years, growing demand for natural gas in the Northeast has led to increased capacity along lines serving both New York and Boston. The completion of additional gas-fired power plants will further expand demand in these areas.*

15,000
12,000
9,000
6,000
3,000

Capacity Cubic Feet per Day (millions)

Natural Gas

Pipeline Delivery Most of the natural gas delivered to New York City begins life in the gas fields of Texas, Louisiana, or western Canada. It generally takes about five days to reach New York, traveling through 42-inch-wide pipes at an average speed of about 15 mph, and enters the city via what is known as the "city gates," a location at which custody of the gas passes from the pipeline company to the local distribution company.

The pipelines serving New York are part of a 250,000-mile interstate transmission pipeline built in the 1950s. To flow effectively over the 1,800 or so miles that it must travel, the gas must be pushed and pressurized numerous times en route. This occurs at compressor stations, located every 40 to 100 miles along the pipeline. Within the station, a turbine will pump up the pressure of the gas, enabling it to continue its journey. As communities tap into the line en route, the pipes often grow smaller—further helping the gas to maintain its pressure.

Four interstate pipeline companies—Transcontinental, Texas Eastern, Tennessee Gas, and Iroquois Gas—serve New York City through seven major interconnections. These companies are essentially transportation providers: like railroads or trucking companies, they do not own the commodity that moves through their system. Instead, they act under contract to gas buyers (generally either local gas distribution companies or independent gas marketers), moving the gas from the producing to the consuming region. The business as a whole is highly regulated: transportation and storage rates are set by the Federal Energy Regulatory Commission, which also requires the pipeline companies to provide open access to any shipper requesting gas transport.

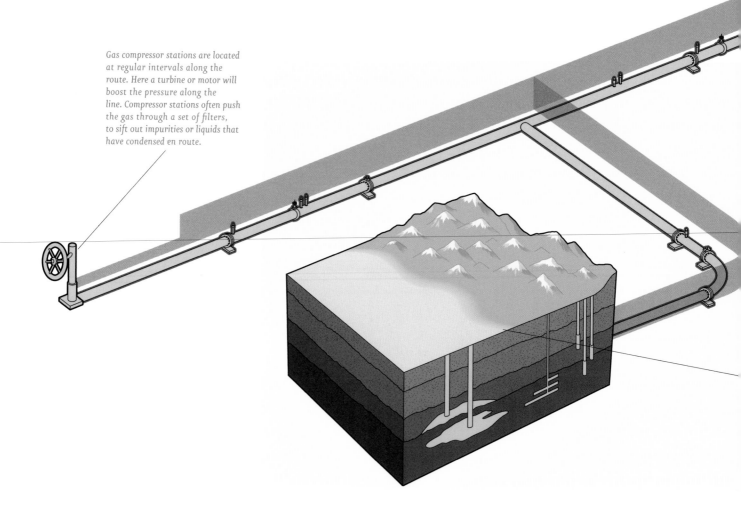

Gas compressor stations are located at regular intervals along the route. Here a turbine or motor will boost the pressure along the line. Compressor stations often push the gas through a set of filters, to sift out impurities or liquids that have condensed en route.

Sections of the pipeline branch off to other gas purchasers (e.g., LIPA or KeySpan) and to in-city power generating stations.

Gas destined for residential and commercial customers moves to an area gas-regulator station, which reduces gas pressure prior to distribution to homes and businesses.

At several locations across the city, the gas pressure is stepped down, an odorant is added for safety, and custody is transferred to Con Edison.

Storing Gas

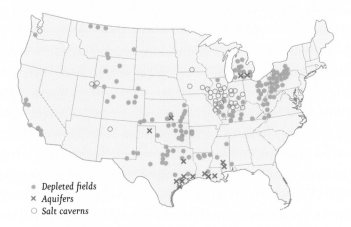

- Depleted fields
- × Aquifers
- ○ Salt caverns

Not all gas travels directly from the gas field to the end user. Large amounts of gas are regularly delivered to storage facilities located in Pennsylvania and western New York State. These storage facilities are generally depleted gas reservoirs, underground caverns created from salt domes, or, occasionally, reconditioned aquifers. Gas delivered to these storage areas will generally complete its journey to the city in the wintertime, when gas demand is highest.

Gas Storage Tanks With the

exception of very limited amounts of liquefied natural gas held at power plants in the city, natural gas is not stored in the city. However, gas storage tanks were once a familiar part of the city's skyline. Two gas storage tanks in Elmhurst, Queens, at 200 feet tall each, were a landmark to commuters on the Long Island Expressway until they were demolished in 1996. An equally famous pair built by Brooklyn Union Gas in Greenpoint, and known as the "Maspeth Holders," were 400 feet tall and held 17 and 15 million pounds of gas respectively. Demolished in 2001, they were known as "Chess and Checkers" for the patterns on their roofs, which were painted in response to a request from the FAA to be "instantly recognizable" from above by air traffic destined for LaGuardia.

Natural Gas

Some pipes are welded or fused together at seams.

Cast iron and metal pipes are often bolted together.

Some pipes are linked together by a special sleeve that holds the two pipe ends.

Local Distribution

An underground complex of gas mains and service pipes brings gas to individual buildings and end users. Just as it is important to keep up pressure in the interstate pipeline, the underground network must also maintain certain pressure levels. Checkpoints are located in manholes throughout the system; these checkpoints feature "regulators," devices that can automatically increase or decrease the flow of gas to meet certain pressure levels. Like compressor stations, regulators are

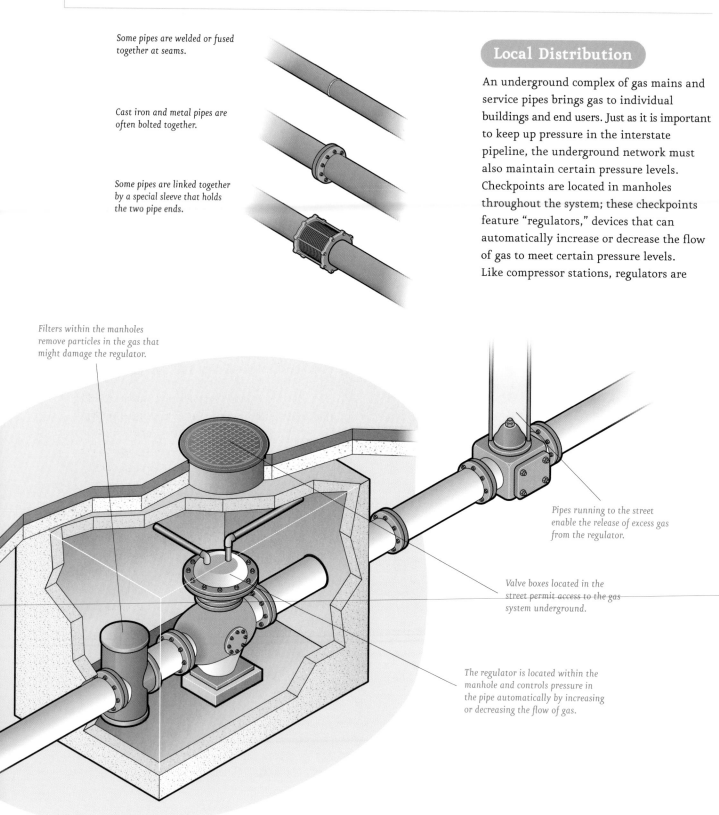

Filters within the manholes remove particles in the gas that might damage the regulator.

Pipes running to the street enable the release of excess gas from the regulator.

Valve boxes located in the street permit access to the gas system underground.

The regulator is located within the manhole and controls pressure in the pipe automatically by increasing or decreasing the flow of gas.

often accompanied by filters that trap any impurities in the flow of gas.

In the early days of New York City's gas network, the mains were made of wood. This practice continued until the 1950s, when cast iron pipes replaced many of the wood ones as an accommodation for natural gas. Today, however, plastic and steel are replacing the cast iron pipes, which were brittle and cracked easily. Most metal and cast iron pipes are bolted, welded, or linked together by a special sleeve that connects the two parts; plastic pipes are generally fused together.

How a Gas Meter Works *A gas meter measures the volume of natural gas that moves through it; it's that simple. The meter contains chambers, each with a specific volume. When an appliance begins operating, gas moves through the meter: first, into a chamber and then out into the appliance itself. The meter simply records the number of times the chamber is filled and emptied. It generally measures gas consumption in thousands of cubic feet.*

Pigging the Pipe In addition to the two traditional methods of detecting leaks, smell and observation of pressure losses, there is now a third—a sophisticated piece of detection equipment known as a "pig." Pigs are robotic devices that are propelled down pipelines to check for signs of corrosion, detect small leaks, and otherwise test the state of the interior of a pipe. Sending a pig down a pipeline is generally referred to as "pigging" the pipe.

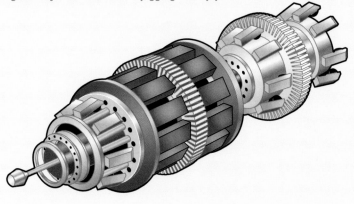

Repair and Maintenance

Gas leaks in a system as old as New York's are not unusual. Because natural gas has no odor, the distributor—in New York's case, Con Edison—adds an odorant to gas, called mercapan, that has the smell of rotten eggs. Smell alone is one aspect of the early warning system for gas leaks; the other is observation of pressure losses, which can be seen from the control rooms of either Con Edison or KeySpan.

To facilitate repairs to leaking gas mains, which are generally located at least three feet underground, shut-off valves and bypass pipes are located throughout the system. Most frequently, repairs involve cutting open trenches to access pipes. More recently, however, Con Edison has experimented with microtunneling— an approach that involves a "Pitmole" device to assist in installing new gas lines with minimal surface disruption.

Every year, some 30 billion pounds of steam flow beneath the streets of Manhattan

from the Battery to 96th St. New York's subterranean steam network constitutes the biggest district steam system in the world—double the size of Paris's, Europe's largest system, and bigger than the next four largest American systems combined.

Steam

From seven generating plants across the city, steam flows at up to 75 miles per hour through over 100 miles of mains and service pipes (and under 1,200 steam manholes) to reach many of New York's most famous institutions. The United Nations, the Metropolitan Museum of Art, and the Empire State Building are all big steam customers. Along with thousands of other households and businesses, these institutions rely on steam for heating, air conditioning, or hot water.

Although impressive in size, New York's steam system is in some ways a relic of a different age. The New York Steam Company made its debut in 1882, offering a technology that would dramatically reduce the amount of soot coming from individual coal-burning furnaces. The basic framework of the system set up by the company—while expanded and improved upon by Con Edison over the years—remains in place; indeed, some of the original pipes, though no longer in use, can still be found under city streets.

The Worthington Steam Pumping Engine was used by several different municipalities, including Brooklyn, to distribute steam to consumers in the late 1880s.

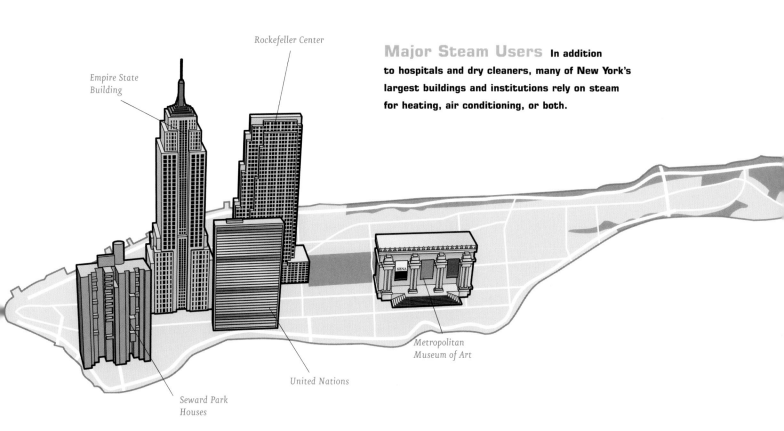

Rockefeller Center

Empire State
Building

Major Steam Users In addition
to hospitals and dry cleaners, many of New York's
largest buildings and institutions rely on steam
for heating, air conditioning, or both.

Metropolitan
Museum of Art

United Nations

Seward Park
Houses

Today, steam represents about 7 percent of Con Edison's revenues. It is a seasonal business, with demand peaking at the coldest times of the year. To meet peak winter demand of up to 12 million pounds per hour, the system consumes roughly 1.6 million gallons of water per hour—making Con Ed the city's largest water user and perhaps its most concerned drought watcher. Summer peak use reaches 8 million pounds per hour; fall and spring demand averages roughly half of that.

Steam customers vary from very big buildings to very small businesses. In all, there are close to 100,000 homes and businesses that rely on the district steam system, many of whom rely on it for more than just heat. For dry cleaners, for example, the steam system is a gift. When employees arrive at the beginning of each day, it takes less than a minute to open a steam valve that supplies steam to the store. Nearly every piece of dry cleaning equipment will make use of the steam: the "spotter" that removes stains, the "puffer" that removes wrinkles, the "presser" that irons clothes, and indeed the dry cleaning machine itself.

Hospitals make regular use of steam as well, mainly for sterilization. At St. Vincent's Hospital, for example, nearly 200 cleaning trays—containing anywhere from six to 130 surgical steel instruments— move through the hospital's Central Sterilization Unit each day. After being washed, they move into a sterilizer—where the intense heat from the steam destroys any bacteria that linger.

Steam

New York City's Steam Network

74th St. Station

from Ravenswood Station, Queens

60th St. Station

59th St. Station

Waterside Station

East River Station

from Hudson Ave. Station, Brooklyn

Distribution Steam is most effective where people are densest. A centralized system—one where only a few large boilers provide heat to many consumers—relies on its ability to save space and cost, to make it worthwhile for the individual customers. In New York, the two most concentrated areas of distribution are, not surprising, where buildings are tallest—in lower Manhattan and in midtown. North of 96th St., the contour of the bedrock flattens, limiting the height of the buildings; it is unlikely that owners of the "shorter" buildings there could ever use enough steam to justify the cost of extending the mains northward.

Adieu to Waterside The oldest steam plant in New York is Waterside, located between 37th and 40th Sts. on 1st Ave. just south of the United Nations. Sited along the river, for easy access by coal barges, Waterside came on line in 1901 and produced steam throughout the twentieth century. It switched fuel twice—first to oil and then to natural gas—and was among the first in-city cogeneration plants, producing steam and electricity simultaneously as early as 1930. Today, Waterside is more valuable as a real estate asset than as a power plant: Con Ed sold it to FSM East River Associates, which plans to develop a mixed-use enclave on that site. While the plant ceased producing steam in 2005, its output will not be missed: Con Ed's East River plant at East 14th St. has been "repowered" with new heat recovery and steam generation equipment, ensuring the system as a whole maintains its current steam capacity.

Steam is produced at seven Con Edison generating plants —five of them in Manhattan and one each in Brooklyn and Queens. Three of these plants are cogeneration plants, producing steam as a by-product of electricity generation. Con Ed also receives steam under contract from a steam plant at the Brooklyn Navy Yard.

At each of these plants, water is heated in gas- or oil-fired boilers to a temperature of roughly 1,000 degrees under very high pressure. These boilers are vast: one of the two boilers at the East River plant on 14th St. and 1st Ave. is nine stories high. Each gallon of water will become eight pounds of steam. The steam leaves the furnace and enters the pipes at about 350 degrees Fahrenheit, with a pressure of about 150 pounds per square inch (psi).

The pipes that carry the steam from the plant may be as large as several feet in diameter; most commonly they are between two and three feet. Most are made of steel, though many of the older cast iron pipes remain in service. These older pipes are vulnerable to cracking, and are often coated in asbestos—which is not problematic so long as they remain undisturbed.

Steam mains generally lie between four and 15 feet below the street surface, though some (including a very deep one under the Metro-North tunnel running along Park Ave.) may be as deep as 30 feet. Seen from above, the network looks something like a gridiron—laid out so that one generating station can provide load to the entire system if needed.

Delivering Steam

Steam vents *Street-level plumes of steam, normally the result of water coming into contact with a steam pipe, are common sights on the streets of Manhattan. But they are also a serious traffic hazard. Where the fog is particularly dense, orange-and-white-striped steam vents are placed above the manhole to carry the steam well above windshield level to maintain visibility for drivers.*

Steam pipes *are generally wrapped in a thick coat of insulation and encased in several inches of concrete to prevent their heat from affecting other underground pipes and wires.*

Valves *are accessible through manholes or boxes located in the street. A long-handled key enables technicians to reach down from the street to the valve to shut off or otherwise adjust a particular steam pipe.*

*Steel straps, known as **anchors**, are welded around the pipe at certain locations to hold it in place. These are embedded in the concrete that encases the pipe.*

Expansion joints, *some 3,000 in number, are located along the steam system to allow for expansion and contraction of steam pipes. They generally feature a bellowslike elbow, which bends to absorb any movement of the steam pipe.*

Steam

Maintenance and Repair   New York's steam system, considering its age, is relatively incident-free. Over the last two decades, only two or three problems have made themselves felt across the system. Having said that, when a steam explosion does occur it can be quite spectacular, involving tremendous damage to the street infrastructure and occasionally causing loss of life.

More routinely, however, the steam system requires an ongoing program of repair and maintenance. Leaks are not uncommon in the century-old system, and fixing them occupies a large body of repair technicians in Con Edison's steam division. It also involves a piece of very modern machinery: the unique "Welding and Inspection Steam Operations Robot," or WISOR, constructed by Honeybee specifically for Con Edison in 2001. Operated via a 200-foot umbilical cord that provides power, air, and signals to the robot and transmits data to its human operator, the WISOR carries out repairs to leaking pipes *in situ*, avoiding the need to cut open the streets above.

The WISOR Robot that **Con Edison** uses to repair steam pipes underground, without tearing up streets, has three parts: a milling section in the front, a midsection that features retractable legs that allow the robot to move along the pipe, and a rear welding section. After identifying a broken pipe joint with one of its four on-board cameras, WISOR cuts a groove and puts down a new weld in the groove.

Steam Explosions The most significant recent explosion in New York's steam system occurred in August of 1989 near Gramercy Park, and killed two steam workers and a neighborhood resident. The cause of that explosion, like many others, was a phenomenon known as "water hammer," which results from the condensation of water inside a steam pipe. In this case, steam personnel had allowed water to accumulate in a pipe they had turned off for service; when the steam valve was reopened the 400-degree steam hit the relatively cool water, with explosive results.

Water condenses and collects in an inactive section of a steam pipe.

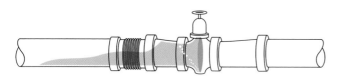

As the line is pressurized and steam is introduced, more condensation takes place.

Once a valve is opened and steam is introduced to the water, bubbles begin to form in the pipe. As more steam is introduced, a large bubble forms near the top of the pipe.

When the bubble collapses, steam rushes into the empty space at 10 times its normal pressure—a level of pressure no main is designed to withstand—and an explosion results.

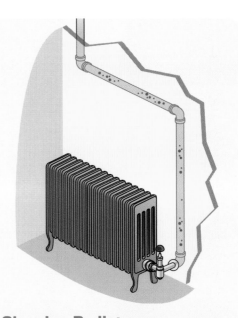

Clanging Radiators Whether their steam comes from Con Ed's central system or from a boiler in their basement, most New Yorkers are all too familiar with banging radiators, though few know what the noise is all about. To reach apartments in a high-rise building or even to reach rooms in a house, steam has to travel through pipes that turn and bend several times over. While this isn't a problem for gas, which is shapeless, it can be a problem for steam, particularly if a few drops of water have condensed inside the building's steam pipes. While the water drops, known as a "traveling slug," can travel as fast as the steam through the system, what they can't do is make the turns in the pipes. When a traveling slug slams into a corner, it produces the harmless—but annoying—banging sound we hear.

COMMUNICATIONS

Cities depend on communications for their livelihood, and New York—as a center of media, of finance, and of advertising—is no exception. Millions of transactions each day move into and out of New York's financial hub, quickly and securely. Beyond the world of Wall Street, the region comprises not only one of the biggest telephone and cell markets in the country, but the largest television and radio market as well.

Not all communications moving into and out of the city are high tech, however. As in most cities, New Yorkers still rely on the postal service for their everyday communications—for letters, for bills, for magazines, and for business correspondence. And despite the proliferation of cable television and satellite, they continue to rely on television and radio transmission—for the news, for entertainment, for traffic and weather.

cellular network. Together, these New York–centric phone calls make up roughly 5 percent of the telephone industry's annual nationwide revenues—a staggering $12.5 billion each year.

As important as the New York market is to the telephone industry nationwide, the importance of the telephone network to the city is even greater. Each day, a trillion dollars of financial transactions move over the city's telephone lines. The

Each day, roughly 125 million telephone calls originate or terminate in New York City. About

half of them travel along the intricate network of copper or optical fiber cables that make up the city's landlines—enough cable, if stretched and laid end to end, to reach the sun. The remainder of the calls rely on the fast-growing wireless or

city's police, fire, and EMS services rely on it heavily for incident management. And there are approximately 10.5 million cell-phone subscribers now in New York City, a testament not only to the convenience of portable phones but also to the real advances in safety, security, and well-being that the cell phone has brought to everyone from school-age children to senior citizens.

But New York's telephone dependence is a fairly recent phenomenon; 100 years ago, telephony was what might be termed an emerging industry. New York Telephone installed its first exchange at 82 Nassau St. in lower Manhattan in 1879. The opening of the exchange led to New York's first "telephone directory"—a card provided to subscribers containing their 252 names. Once Bell Telephone's second patent expired in 1894, independent telephone companies by the hundreds were formed across the country. Within just a few years, more than a dozen companies had been granted licenses to provide phone services within the city.

The multiplicity of phone companies created interconnection problems, however, and subscribers from one company could

Telephone

Before New York City buried its cables underground following the Blizzard of 1888, an intricate network of overhead wires carried its calls from person to person.

not always reach those of another. In 1913, AT&T agreed to connect noncompeting independent telephone companies to its network and a long-distance monopoly was born. It also controlled the 22 Bell operating companies, which provided local telephone service to most of the United States in competition with a handful of smaller providers. For the next 70 years, until its breakup by the federal government in 1984, the Bell system would function as a regulated monopoly, serving New York—and the rest of the country—extremely well.

Today, an estimated six million telephones can be found in the city—and all rely on the technology discovered over a century ago: the sending and receiving of voice messages that have been converted into electrical impulses. The largest player in the local telephone market is Verizon, the company formed by the merger of Bell Atlantic and GTE in 1998. Verizon has the most customers and revenues in both line and wireless markets in the entire nation; it is also one of the largest long-distance carriers and the largest directory publisher in the world.

Telephone Elements

An electronic microphone amplifies the user's voice and sends it out as a series of electrical impulses.

A switch is used to connect and disconnect the phone from the network. This switch is often referred to as the "hook switch."

A touchtone keypad and frequency generator are used to signal the particular circuit that needs to be opened to begin a call.

In many phones, a device known as a "duplex coil" blocks the sounds of the user's voice from reaching back to his or her ear.

Modern telephones include a ringer, often comprised of a speaker and circuit rather than the traditional mechanical bell.

The Empire City Subway Company

The first underground telephone cables shared vaults with neighboring power lines, but became noisy with interference from the electricity that traveled through them. To solve the problem, in 1891 a new company—The Empire City Subway Company—was given the exclusive right to build and lease underground infrastructure for all communications services in Manhattan.

The company retains the right today, at least in Manhattan and parts of the Bronx (in other boroughs, companies can build their own pipes). It employs over 300 people and owns roughly 11,000 manholes and 58 million feet of varied conduit—plastic, concrete, vitrified clay, creosoted wood, and iron. A subsidiary of Verizon, it rents space primarily to telecom and cable television providers and carries everything from traditional phone cables to signaling for fire alarms, traffic lights, and police call boxes.

Telephone

Tracking the Path of a Call

1. After a call is placed, the voice data travel along a pair of copper wires to a local networked phone company box. This box contains hundreds of wire pairs.

2. The call is then routed to a "digital concentrator," where it is digitized at a sample rate of 8,000 samples per second.

3. The digitized information is then sent—usually via coaxial or fiber optic cable—to the phone company's switching station, where it will be routed to its destination.

4. If the call is a local call, the signal is simply looped back into the local system. If it is a long-distance call, the signal is forwarded to the long-distance network.

5. Long-distance calls may be transmitted by cable, microwave towers, or satellite.

140 West St. *Among the most notable of New York's switching stations is 140 West St., a 31-story facility constructed in 1926 as the headquarters of the original New York Telephone Company and more recently Verizon's central office. Made of steel, brick, and stone, it survived the collapse of its neighbor, 7 World Trade Center, on 9/11.*

33 Thomas St. *The 29-story Long Lines Building (Long Lines was once the name of the AT&T division that operated the long-distance toll network) stands sentry in lower Manhattan. The building, built in 1974, is said to be able to withstand nuclear fallout and has enough backup energy to operate for two full weeks.*

811 Tenth Ave. *Like other switching stations, the function of the AT&T switching station in Clinton has determined its shape.*

375 Pearl St. *The New York Telephone Company switching station on Pearl St., adjacent to the Brooklyn Bridge, is clad in white marble slabs with windows represented by black vertical stripes.*

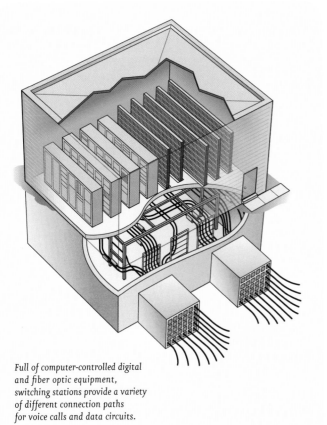

Full of computer-controlled digital and fiber optic equipment, switching stations provide a variety of different connection paths for voice calls and data circuits.

Switching Stations Switching stations are the lynchpin of telephone communications, and there are 80 or so of them spread out across the five boroughs. Each switching center contains a complex array of computer-controlled digital and fiber optic equipment that fulfills a variety of functions: call processing, billing, distribution, and enhanced features for subscribers. Each also contains sophisticated power, ventilation, and cooling systems to ensure the continued operation of the equipment. One feature absent from many switching stations is windows: the absence of glass panes helps to protect telephone switches and sensitive electronic equipment from dust, temperature, and humidity.

In addition to switching stations, New York's telecommunications infrastructure includes carrier hotels, which locate network carriers and service providers together to allow direct connections between them and to facilitate access to multiple local and long-distance networks. Perhaps the premier carrier hotel in the world is located at 60 Hudson St. in Manhattan, the former Western Union headquarters. Over 100 domestic and international telecom companies are housed there, including AT&T, Cable & Wireless, GTE-Verizon, Time Warner Telecom, Qwest, and Global Crossing.

Telephone

Underground Cables

Three types of cables carry telephone signals under the city's streets today. The oldest are lead-covered copper cables; these have been replaced in many places by copper cables covered in plastic. More powerful optical fiber cables are gradually replacing both types of copper cables as the preferred way to carry telephone signals. But the replacement of the old copper cables will not happen overnight: copper wire and optical fiber cable cannot be spliced together, so optical fiber is currently laid only where whole segments of the underground network (from one substation to another) can be replaced.

The wires inside both copper and optical fiber cables are typically found in pairs: one for the outgoing message and one for the response. But the optical fiber cables are lighter, less bulky, and immune from disturbance by moisture or nearby electrical currents. They also provide far more capacity. A typical copper cable can transmit about 25 conversations in analog mode, while a single optical fiber, slightly wider than a human hair, can typically transmit over 193,000 conversations digitally.

How Optical Fiber Works

Optical fiber cables are thin strands of glass that convey information via light pulses generated by lasers. To send telephone conversations through a fiber optic line, traditional analog voice signals are translated first into digital signals (1s and 0s, represented by on/off light beams). A laser at one end of the pipe switches on and off to send each bit of data—billions of times per second. Multiple lasers with different colors are now commonly used to fit multiple signals into the same fiber.

The light in the fiber optic cable will then travel through the core by bouncing off the "cladding," or mirror-lined walls. Because the cladding does not absorb any light moving through the core, the light wave can travel a long distance without losing strength. Fiber optic cables can easily carry signals 50 miles or more. To move farther than that, the system relies on an "equipment hut," where a machine picks up and retransmits the signal down the next segment of line at full strength.

Copper cables are made of approximately 2,700 wire pairs grouped in clusters and bound with different color-coded wires for easier repair identification. Cables are about three inches thick and are insulated with aluminum and neoprene.

The Mystery of the Silver Tanks

Shiny silver tanks have become an oddly familiar site on New York streets, particularly in lower Manhattan. These 40-gallon nitrogen tanks are used by Verizon to prevent moisture (from steam and heating systems) from damaging underground phone lines. Each tank stores nitrogen in liquid form at –300 degrees Fahrenheit and delivers it belowground via a small rubber tube running through a manhole cover. As it is released, the nitrogen heats up to –280 degrees and becomes gas, providing pressure that travels along the voice and data cables and keeps them dry. The tanks, roughly 50 of which can be found on city streets at any one time, last only three days before they must be replaced.

A Faulty Line

While work is under way to repair the line, fresh air is pumped into the manhole through a large sleevelike tube. A steel collar stands in the manhole opening to prevent anything from falling into the hole while repair work is in progress.

Once work is completed, the cable is placed inside a plastic or lead sleeve, which is kept under pressure and sealed. Any break in the sleeve will result in a reduction in pressure and trigger an alarm.

If a faulty line needs to be repaired, a repairman will look up the number of the cable through which the wire passes. This will direct him to the appropriate manhole and, once inside, to the correct cable.

To facilitate repair, cables are hung in brackets or racks within the manhole. Cables are about three inches in diameter, and the wires inside are grouped in clusters and bound with colored wire for identification purposes.

Boosting a Call *All signals, whether carried by copper or fiber optic cable, degrade over distance. However, because fiber optic cable conducts signals better, boosters for these cables can be placed farther apart than those for copper cables.*

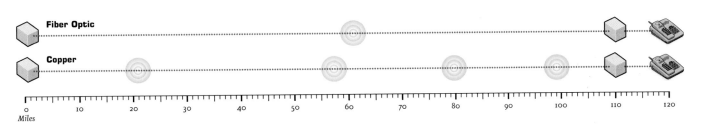

Fiber Optic

Copper

0   10   20   30   40   50   60   70   80   90   100   110   120
Miles

Telephone

Cell Phones New York City is the largest cellular market in the country, with nearly half of all calls made within the city today carried on a cellular network. And that number is growing rapidly: the number of minutes of cell-phone calls within New York State is projected to increase by 37 percent over the next three years, while the number of minutes of landline calls is projected to decrease by 8.4 percent. But popularity should not be confused with level of service: various surveys have rated New York as among the worst metropolitan areas in the country, in terms of blocked and dropped calls, for cellular coverage.

For even the most experienced wireless carriers, serving the New York market is a tricky undertaking. Cell signals are subject to obstruction from tall buildings, of which there are many; these "dead zones" are not always easily brought back to life. The system can also fail due to overload; with millions of registered cell phones in the city, "dropped" calls are common at peak calling hours (such as 5 p.m.). While these dropped calls are generally more an inconvenience than anything else, they can at times prove serious: an estimated 120,000 calls to 911 did not go through in 2002 due to cell-phone failure.

How Cell Phones Work

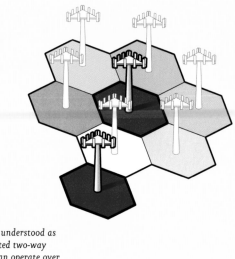

Cell phones are best understood as extremely sophisticated two-way radios—ones that can operate over multiple channels within the same geographic area. Each geographic area is divided into cells, which are allocated a set of voice channels.

The Journey of a Cell Call

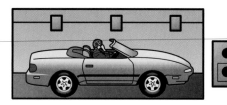

A call is placed from a car inside the Holland Tunnel. This call is sent by an antenna—called a radiating cable—to a computer at the mouth of the tunnel.

The call is then routed to the sender's service provider, which sends it to a computerized center that handles many cell sites.

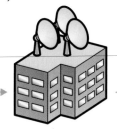

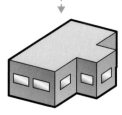

If the call is destined for a cell phone serviced by another provider or for a landline, it will be sent into the local telephone network and treated like a call made on a landline.

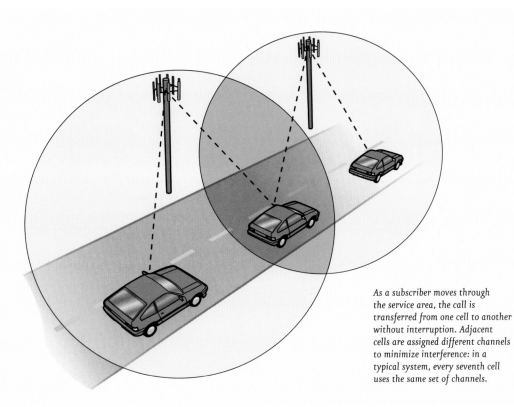

As a subscriber moves through the service area, the call is transferred from one cell to another without interruption. Adjacent cells are assigned different channels to minimize interference: in a typical system, every seventh cell uses the same set of channels.

If the call is destined for a phone served by the same provider, it is then transmitted to an antenna near the recipient.

A radio wave sends the call from the antenna to the call's recipient.

The call may then pass through several routing facilities until it reaches the routing center of the recipient's service provider.

Telephone

Pay Phones Roughly half of all public pay telephones (PPTs) in the city are located on city sidewalks and are regulated by the Department of Information Technology and Telecommunications (DoITT). The department does not handle phones located within the subway system (handled by the MTA), those inside city parks (handled by the Parks Department) and those inside private property (e.g., gas stations, hospitals, and office buildings). For regulatory purposes, there are two types of sidewalk phones: those located within six feet of the building line (building line PPTs) and those beyond the six-foot line (curbside PPTs). Only the latter is permitted to carry advertising, and that advertising is restricted to contracts with one of four designated media representatives.

Sixty-three different companies are responsible for the 30,000 public pay telephones regulated by DoITT. Each of these companies has been granted a franchise by the city—the nonexclusive right to install, operate, and maintain a pay telephone on the city's sidewalk. In addition to the franchise, each pay phone must pass certain siting criteria relating to pedestrian movement and street furniture spacing. In addition to meeting DoITT's criteria, the relevant community board and the building landlord (if it is a building line PPT) must be notified. The phone is inspected by DoITT upon installation to ensure that it meets specifications; it will also be inspected subsequently to ensure it remains in working order and is cleaned regularly (twice a month).

Citywide Pay Phone Distribution

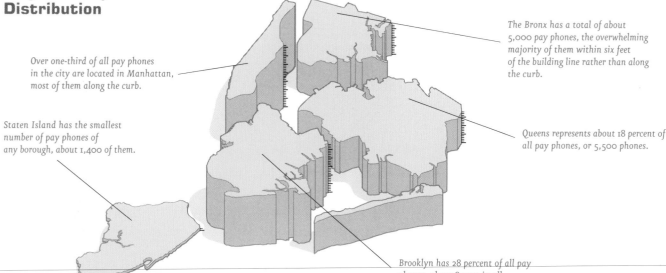

Over one-third of all pay phones in the city are located in Manhattan, most of them along the curb.

Staten Island has the smallest number of pay phones of any borough, about 1,400 of them.

The Bronx has a total of about 5,000 pay phones, the overwhelming majority of them within six feet of the building line rather than along the curb.

Queens represents about 18 percent of all pay phones, or 5,500 phones.

Brooklyn has 28 percent of all pay phones, about 8,000 in all.

Pay Phones Galore As of spring 2004, there were just under 30,000 active public pay phones on the city's sidewalks; another 30,000 or so can be found in private premises. Although pay phone usage has decreased as the number of cell phones has increased, there is still a pressing need for these phones. Cell-phone batteries die, cell-phone service may be interrupted, and an estimated 20 percent of the city's citizens do not have regular phone service in their homes.

N11 codes N11 codes allow callers to dial three digits, instead of seven or 10, to connect to a certain telephone destination. The network is preprogrammed to translate the three-digit codes (e.g., 911) into the relevant seven or 10-digit telephone number and to route the call accordingly. Generally, the first digit can be any number other than 1 or 0, while the last two digits are both 1.

211: Community information and referral services

311: Nonemergency police and government services

411: Unassigned, but used nationwide by carriers for directory assistance

511: Traffic and transportation information

611: Unassigned, but often used by carriers for repair services

711: Access to telecom relay services (assigned nationwide)

811: Unassigned, but may become the national number to call before streets are dug up for utility or other work

911: Unassigned, but used nationwide for emergency services

311 Perhaps no innovation in the world of urban telecommunications has been appreciated by more New Yorkers than the introduction of the 311 call line in March 2003. Designed to replace over 40 call centers and help lines in dozens of agencies across the city, the 311 call center—based at Maiden Lane in lower Manhattan—operates 24 hours a day, seven days a week as a one-stop shop for civic gripes. Some 250 agents speaking 170 different languages answer citizens' questions about government services—anything from reporting potholes to broken streetlights to air conditioner disposal. Service requests, such as noise complaints, are electronically transferred to the relevant city agency—in this case the Police Department—for action.

Currently, 311 operators handle about 40,000 to 45,000 calls each day. All of them are answered by a live operator, and quickly. Roughly 95 percent of calls are answered within 30 seconds; the average wait time is an impressive seven seconds. Interestingly enough, during the first year of 311 operation, the 911 emergency network experienced a quarter of a million fewer calls than the previous year—the first decrease in 911 call volume in 13 years.

At the city's 311 call center, on Maiden Lane in lower Manhattan, up to 250 call takers respond to questions on every aspect of city life.

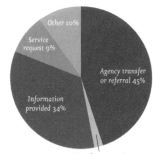

Other 10%

Service request 9%

Agency transfer or referral 45%

Information provided 34%

Transfer to 911 2%

311 Inquiries Handled Although many 311 inquiries are seasonal, almost half of the center's incoming calls are for specific agency transfers or referrals.

Daily 311 Call Volume Trend and Service Levels

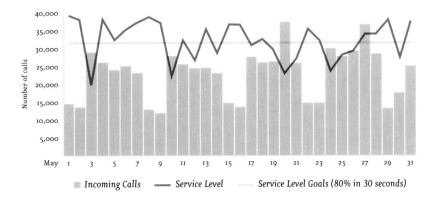

Number of calls

40,000
35,000
30,000
25,000
20,000
15,000
10,000
5,000

May 1 3 5 7 9 11 13 15 17 19 21 23 25 27 29 31

▪ Incoming Calls — Service Level Service Level Goals (80% in 30 seconds)

The 311 center keeps track of the number of calls on an hourly, monthly, and annual basis. It also keeps track of whether more or less than 80 percent of the calls are answered within 30 seconds. The city also subsequently analyzes how the various agencies respond to the forwarded inquiries.

Telephone

Emergency Communications Nearly every New Yorker knows the emergency telephone number: 911. What many do not know is that this number is the emergency number for more than New York: since the early 1980s, it has been the emergency number for the entire country. In over 99 percent of locations in the United States and Canada, dialing 911 from any telephone will generally link the caller to an emergency dispatch center, which will send relevant emergency responders to the caller's location. The one exception to this is mobile or cellular phones: in certain locations, these calls will go through to the state police or highway patrol (rather than the local authorities), who will transfer the call to the appropriate local emergency services once the caller has described his or her location.

In New York City, residents dial 911 an average of 23 times each minute—or close to 12 million times a year. Operators who take the calls work for the Police Department. The costs of the system are largely borne by telephone users: since 1992, the 911 system has been funded by a state surcharge on phone bills. Originally 35 cents, the tax was raised to $1 in 2002. Cell-phone customers also pay a tax: they are charged $1.20 each month for "enhanced 911," an upgraded service being developed that will enable the location of callers to be traced, as well as a 30-cent local surcharge "for purposes of enhancing the public safety communications network."

Responses to 911 within the city are not always seamless: eight dispatching centers are ready to respond but are not always working on the same information. Police computers, for example, automatically display the addresses of 911 callers—but fire dispatchers have no way of seeing that information. The centers are also physically and

The 911 System, Today and Tomorrow

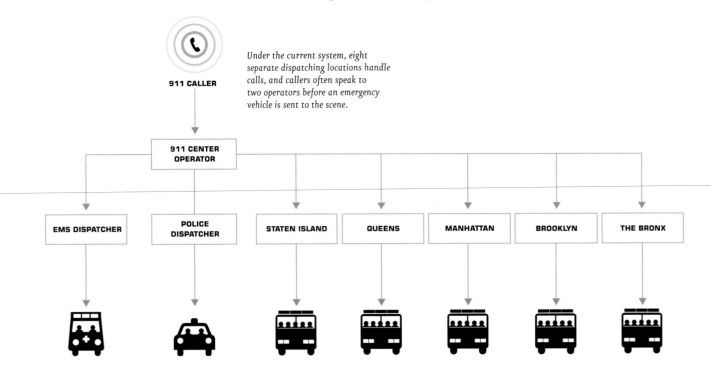

911 CALLER

Under the current system, eight separate dispatching locations handle calls, and callers often speak to two operators before an emergency vehicle is sent to the scene.

911 CENTER OPERATOR

| EMS DISPATCHER | POLICE DISPATCHER | STATEN ISLAND | QUEENS | MANHATTAN | BROOKLYN | THE BRONX |

technologically isolated from one another: oftentimes they have to dial back into the 911 system to share information across agency lines. Even operations within the same organizational umbrella are not located together: although the Fire Department took control of the Emergency Medical Service (EMS) in 1996, the two agencies continue to work out of different dispatch centers and on different computers.

All this will change if current plans to integrate the 911 service come to fruition. The plan involves uniting the disparate computer systems and dispatching centers into two identical dispatching centers relying on one unified computer system. (A similar system exists in Chicago, where all dispatchers work in one physical location and information can be displayed quickly in police and fire vehicles.) However, significant technical and political obstacles suggest that the long-awaited unification may be some years away.

Fire Alarm Boxes are a familiar sight on many city streets, located either on poles or on stand-alone, decorative pedestals. These boxes date back to 1870, when the Fire Department installed fire alarm boxes on telegraph poles south of 14th St. Though few boxes from that time have survived, a large number of those on the streets today still rely on the original technology: pulling a revolving coded-wheel mechanism sends a signal identifying the box number to the central office of the borough in which it is located, and dispatchers there forward the alarm to the appropriate firehouse. (More recent models feature a speaker, which allows the caller to contact the Police or Fire departments directly.) Although the proliferation of cell phones has meant that wholesale reliance on the 911 call-in system makes more sense, there has been localized opposition to the idea of removing fire alarm boxes and the city continues to be required to maintain them.

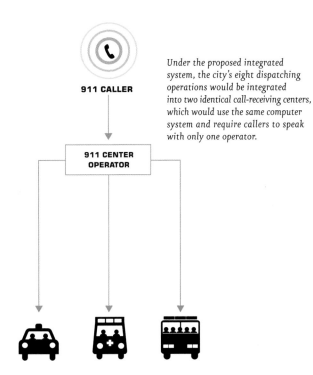

911 CALLER

911 CENTER OPERATOR

Under the proposed integrated system, the city's eight dispatching operations would be integrated into two identical call-receiving centers, which would use the same computer system and require callers to speak with only one operator.

among the largest of its kind anywhere in the country.

The first official notice of a postal service in New York dates back to 1673, when a monthly postal service between New York and Boston was initiated. The service lasted only a short time, but the post rider's trail came to be known as the Boston Post Road, which forms part of what we know today as U.S. Route 1. Street boxes for mail delivery did not appear until well over a century later, in the mid-1800s. Free delivery was provided to 49 of the largest U.S. cities—including New York—in 1863; not until the turn of the last century would it be provided free to rural dwellers.

Each day, roughly 23 million pieces of mail are processed in New York City—

most of them at night. In addition to 263 post offices spread across the five boroughs, the United States Postal Service (USPS) maintains five processing centers in the city—one in each borough and an additional one at JFK Airport. The busiest, the 2.2 million-square-foot Morgan Processing and Distribution Center in Manhattan, is

Horse-drawn wagons and carriers on foot provided the primary means of delivering the mail through most of the nineteenth century. By the end of the century, however, engineers had devised a way to move the mail in the nation's larger cities: the underground pneumatic mail tube. First installed in Philadelphia in 1893, the tube system featured a two-foot steel cylinder that was driven through a cast iron pipe pressurized by electrically driven air compressors and rotary blowers. Each eight-inch canister was designed to hold up to 500 letters and could travel up to 30 miles per hour.

Moving the Mail

In New York, the first pneumatic tube was put into operation between the General Post Office and the Produce Exchange in 1897 by the American Pneumatic Service Company. It was soon extended to Grand Central Station, northward to Harlem, and over the Brooklyn Bridge to the General Post Office in Brooklyn. In total, New York's pneumatic system traversed a 27-mile route and connected 23 post offices. Each route featured two pipes: one receiving and one for sending. At its peak, roughly one-third of all first-class letters that moved

The old post office (circa 1890) was located at the southern end of City Hall Park.

through New York's main post office were distributed to branch offices by pneumatic tubes.

The high operating costs of the pneumatic system ultimately proved its downfall. By 1918, the federal government considered the annual rental payments ($17,000 per mile per annum) made by the post office to be "exorbitant" and endorsed a new alternative with greater capacity—the automobile— as the delivery method of choice. Chicago, Philadelphia, and St. Louis lost their service first. New York halted use of the tubes temporarily, but successful lobbying by contractors led to resumed service in 1922. The system continued in use until 1953.

Pneumatic Facts

Hours of operation: 5 a.m. to 10 p.m. weekdays and Saturday 5 a.m. to 10 a.m.

Rate: 5 tube canisters per minute

Speed: 30 miles per hour

Pressure: 3 to 8 pounds per inch

Size of carrier: 24 inches long, 8 inches across

System capacity: 400 to 500 letters

System mileage: 26.96 miles of two-way tubes

Number: approximately 95,000 letters daily

How the tube worked The two-foot-long canister had felt and leather packing on each end to create an airtight seal. It also had four small wheels, to prevent it getting stuck at junctions in the pipes. Perforated steel cylinders filled with oil were sent through the tubes to lubricate them to assist the passage of the canisters.

The Pneumatic Tube Mail Network

The two northernmost stations in the network were Manhattanville and Triborough, and it took between 15 and 20 minutes for letters to get there from Herald Square.

It took about 11 minutes to get from the General Post Office at Herald Square to the Planetarium Post Office near the Museum of Natural History.

It took only four minutes to get from the General Post Office at Herald Square to Grand Central.

It took four minutes to cross under the East River and move a letter from the Church St. Post Office to the General Post Office in Brooklyn.

The pneumatic system was located four to six feet below city streets. It formed a loop running north from City Hall, with an extension from there into Brooklyn.

Moving the Mail

Mail Distribution The postal system in New York City is a vast operation, involving over 20,000 employees spread across post offices and processing centers. Notwithstanding this enormous labor force, it is rare that any of them would ever touch a letter dropped in the local mailbox. Nearly 90 percent of the processing operation is automated—human intervention is largely reserved for wrong or unintelligible addresses, or disintegrating letters.

Once a letter is picked up, it moves either through a local branch office or directly to one of the five city processing centers, where it travels through a sorting and stamping machine. With the help of an ultraviolet light that picks up phosphorescent dye in the stamp, the letter is canceled, boxed with others, and moved to a letter-sorting machine. Traveling on a long conveyor, the letter will be sorted by zip code and bundled with others for distribution within the same geographic destination.

From the sorting office, letters are readied for delivery either by surface or by air. Those not destined locally or for a nearby state are taken by truck to one of the region's three airports. All three are equipped with postal facilities where mail is bundled and turned over to the airlines for loading on the appropriate commercial plane. Theoretically, mail traveling locally should arrive the next day, mail going to neighboring states is expected to take two days, and service from the East to West coasts should take three.

A Step-by-Step Guide to Mail Delivery

A letter carrier gathers mail from a collection box on the street.

The mail is brought by a mail van to a local branch office, where it is "rough culled"—i.e., parcels are separated from letters.

From the local branch, the mail is loaded onto a larger, seven-ton truck for delivery to one of the processing centers in the city.

Letter carriers collect their mail, already sorted into walk sequence, and deliver it to homes in the area.

Mail moves from the processing station back to the local post office, where the letters are unloaded and readied for pickup by the carriers.

A second bar-code sorter machine separates the mail into the dozens of carrier routes within a given post office delivery area and then into the order in which it will be delivered on that route.

Relaying the Mail Familiar destinations to most New Yorkers, over 8,000 blue mail collection boxes are spread out across the city. Less familiar are the olive green "relay boxes," which are used as temporary holding spots for letter carriers whose carts or mail sacks cannot hold all the mail to be delivered within their territory. Presorted mail is delivered by postal van to the relay boxes at appropriate points during the day for pickup by letter carriers when they are ready for the next load.

Letters that can't be delivered and don't have a return address may be sent to a "mail recovery center" in Minneapolis—this is often referred to as "the dead letter office." Here, in what's called the "nixie" section, a damaged letter will be covered in plastic wrapping. Mail that is not able to be deciphered locally due to a partial or incomplete address, such as "Police Chief, Kansas City," will be addressed in full and forwarded for delivery. Roughly 30 wallets per day, along with miscellaneous keys and glasses, are dealt with locally.

Conveyors move handwritten and typed letters to a multiline optical character reader, which "reads" the faces of letters and imprints a nine-digit bar code. These readers can process up to 13 pieces of mail per second.

At the processing center, hampers of mail are emptied onto a moving conveyor belt. The belt passes through a series of chutes that segregate thick or oversized mail from letters that will be read by a machine.

Letters proceed to the "advanced facer-canceling machine," which faces all the letters in one direction, seeks out the stamps, and cancels them. The machine also separates the mail into three categories: handwritten, typed, and "previously bar-coded by the customer."

To deal with addresses that cannot be recognized by the advanced facer-canceling machine, the post office relies on the Remote Bar Coding System (RBCS), which takes video images of an address from an envelope. An image will first be sent to an in-house remote computer reader, which tries to match the video "snapshot" with addresses on file.

The bar-code sorter reads the bar codes—40,000 an hour—and separates the mail by zip codes.

Trucks are loaded at the processing center for destinations throughout the region or for the airport. Upon arrival at a local processing center, the mail is weighed and delivered to a bar-code sorter.

Information from letters processed via video image will then be put into a bar-code sorter.

If no match is made, the video image is transmitted to a facility outside of the city, where a postal operator located there will read the address, identify the appropriate zip code, and send the information back to the processing location.

Moving the Mail

Zip Codes As most people know, zip codes are a national—rather than a local—phenomenon. First introduced by the U.S. Postal Service in 1963 as the Zone Improvement Plan (ZIP) code, the system today continues to rely on the familiar five-digit code to speed delivery of mail to addresses across the country.

Each digit of a zip code represents something specific to the post office. The first digit indicates one of 10 large geographic areas of the country, ranging from 0 in the Northeast and 1 in New York to 9 in California and the West Coast. The second and third digits represent urban areas and centers having common transportation access. The last two digits of the zip code indicate local post offices or postal zones in the larger cities.

The advent of the zip code revolutionized mail processing: by making it easy for addresses to be converted to bar codes, it did away with rounds of highly repetitive sorting done by postal workers. Bar-coded mail could be put through a sorter machine, which automatically classifies the mail by destination town and street.

In 1983, the delivery of the nation's mail was further automated with the introduction of ZIP+4 codes. The additional four digits identify more precisely a letter's destination. The sixth and seventh digits indicate a "delivery

Neither Snow Nor Rain... The James Farley Post Office building, on Eighth Ave. between 31st and 33rd Sts., is perhaps the most recognized post office in the city. Its inscription, *"Neither snow nor rain nor heat nor gloom of night stays these couriers from the swift completion of their appointed rounds,"* is not the official motto of the postal service, but rather a sentence taken from the Greek historian Herodotus' works that refers to the system of mounted postal couriers maintained by the Persians roughly 500 years before the birth of Christ.

While the inscription will remain thanks to the building's status as a National Historic Landmark, only a small part of the building will continue to serve as a postal facility in the future. In 2002, the building was transferred to the Pennsylvania Station Redevelopment Corporation, a subsidiary of the state's Empire State Development Corporation, and proposals for redevelopment are being considered. The postal service will retain just over 200,000 square feet of the 1.5-million-square-foot building, primarily for Express Mail, truck platforms, and a stamp depository. Retail lobby services will also continue to be provided at the site.

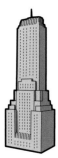

Chanin Building
10168

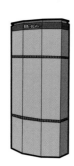

Met Life Building
10166

Seagram Building
10152

Helmsley Building
10169

Park Avenue Plaza
10055

sector," such as several blocks, streets, or office buildings: the eighth and ninth represent a particular delivery segment, such as a floor of an office building or one side of a street between intersections.

Although rare in other places, it is not uncommon for buildings in New York to have their own zip codes. The World Trade Center, for example, carried the 10048 zip code: its 16,000 addresses received on average about 85,000 pieces of mail each day. Currently, 44 buildings in New York City—from the Empire State Building to the Woolworth Building—are large enough to carry their own zip codes.

Zip Codes by the Dozen

The City of New York is served by over 100 unique zip codes and dozens of individual post offices, not to mention 44 buildings in Manhattan with their own assigned zip codes.

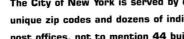

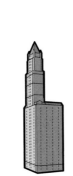

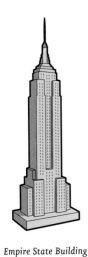

Empire State Building
10118

Woolworth Building
10279

World Financial Center
10281

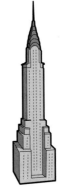

Chrysler Building
10174

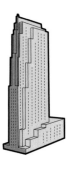

*GE Building/
30 Rockefeller Center*
10112

The Airwaves

Like any large American city, much of New York's communication takes place not underground

but via the airwaves. Also known as the "electromagnetic spectrum," the airwaves are relied upon for an ever-expanding list of creature comforts: radio, broadcast television, cell phones, wireless Internet, car alarms, garage door openers—to name but a few. To protect against interference, each use is assigned a slightly different set of airwaves.

The spectrum is essentially a way to characterize radio waves, which are transmitted at different frequencies (measured in hertz, or Hz, and megahertz, or MHz) and with different wavelengths (a wavelength is the distance between the top or bottom of a wave). The size of the wavelength affects the ability of the wave to pass through objects: as a wavelength decreases in size (and increases in frequency) its ability to move through obstacles like walls or even storms decreases. As a result, higher frequencies are less valuable than lower ones. Popular consumer services like broadcasting and cell phones need to penetrate buildings and therefore need to secure low frequencies. As a whole, the value of the spectrum is estimated at roughly $780 billion, though some estimates—based on recent spectrum auctions—are far greater.

The assignment of spectrum—e.g., car alarms at 300 MHz, or medical implants at 400 MHz—is not done locally, but rather by the federal government. Since 1934, when Congress passed the Communications Act of 1934 in response to a spate of maritime crashes, the Federal Communications Commission (FCC) has assigned and regulated the airwaves. Licensing and coordination on the part of the FCC varies across different "bands" of the electromagnetic spectrum—for example, amateur, cellular, paging, broadband, etc. The federal government leaves some discretion to local officials to coordinate spectrum in the 800 MHz band for public safety, but they must submit their plans for that part of the spectrum to the FCC.

The Radio Corporation of America, forerunner of NBC, broadcast early radio programs from its WJZ studios in New York.

For New Yorkers, with the possible exception of their cell phones, no use of the airwaves is more a part of daily life than radio and television. New York supports roughly 70 radio stations and is generally considered the nation's largest metropolitan radio market, with a potential audience of over 15 million listeners spread across the five boroughs, Westchester, and parts of New Jersey and Long Island. Los Angeles comes next, with a radio market of just under 11 million listeners.

New York has always been an equally important market for the television business. As far back as 1948, there were more households with television in the New York City metropolitan area than in Philadelphia, Los Angeles, and Chicago combined. In the fifties, CBS estimated that one in five of all viewers in the country was located in the New York area. Today, there are an estimated seven million television households in New York, making it by far the most populous market in the nation. Together with Los Angeles, which has just over five million households, the two cities account for about 12 percent of the national television audience.

Rockefeller Center, Then and Now

Perhaps no building in New York is more closely affiliated with radio and television history than 30 Rockefeller Plaza, at Rockefeller Center. NBC began network radio operations there in 1933, with 22 studios and five audition rooms, as well as client and observation rooms, switching booths, a master control center, and technical facilities. The facility simultaneously fed two networks and two local stations.

By this time, NBC had purchased station W2XBS—an experimental television station—from RCA. The station began the industry's first regular schedule of television service on April 30, 1939. Two years later, in 1941, commercial television was initiated by W2XBS's successor, WNBT, with the broadcast of *Truth or Consequences* and *Uncle Jim's Question Bee.* Bulova and Procter & Gamble were among the first sponsors.

Today, Rockefeller Center continues to be home to NBC, now a subsidiary of General Electric. And at 30 Rockefeller Plaza, and in the surrounding complex, are found studios producing some of the nation's most popular entertainment television, including *Saturday Night Live* and *Late Night with Conan O'Brien.* A number of widely watched news programs are also broadcast from here, including *NBC Nightly News* and *Today.*

Anatomy of the Spectrum

Commonly called airwaves or radio waves, the electromagnetic spectrum hosts a variety of everyday communications devices, all operating at different frequencies.

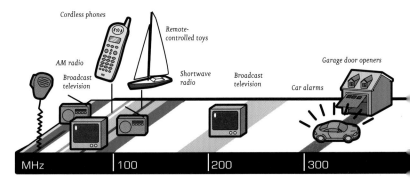

The Airwaves

Television Broadcast television is among the most familiar uses of the airwaves. Most of New York City continues to receive dozens of local terrestrial channels at no cost; residents of the boroughs, depending on where they are located, also commonly receive channels broadcast from farther afield in parts of New Jersey, Connecticut, or Long Island. Despite the proliferation of cable and cable channels, viewership remains high: many more people in New York watch local news on a major network than watch its cable counterpart.

New York is more than just a big television market—it is also an important one. All three of the major networks—the Columbia Broadcasting System (CBS), the National Broadcasting Company (NBC), and the American Broadcasting Corporation (ABC)—have their headquarters in New York: CBS at Black Rock on West 57th St., ABC at Lincoln Center, and NBC at Rockefeller Center. And, equally significant, all three of them broadcast their signature news broadcast—the evening news—from New York.

Delivering the National News
The national news is delivered each night from ABC's studios on West 66th St. in Manhattan. It is distributed via cable to its local affiliate for distribution in the New York area and simultaneously via satellite to local affiliates in other areas of the country.

Local version (for local viewers):
At ABC Television's network building on West 66th St. in New York, coaxial cable carries the television signal to WABC, the local affiliate.

From WABC, the signal is carried via cable to the following:
  - *the local cable supplier's head end, such as Time Warner on 23rd St.*
  - *the Empire State Building, which then transmits a broadcast signal over the air.*
  - *a DTV distribution center, which transmits the signal via satellite.*

Coaxial Cable

ABC TV Network

WABC Local Affiliate

Fiber

Fiber

Fiber

Cable Head End

Empire State Building

Over-the-Air Transmission (NTSC & DTV)

Cable

Home

Home

Direct TV Head End

Home

Direct TV Satellite

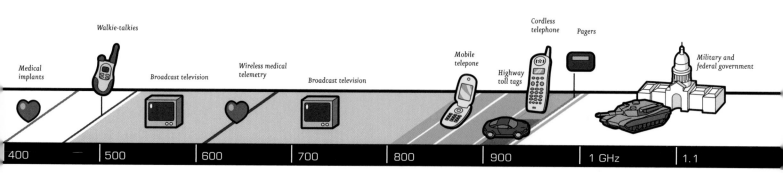

Medical implants

Walkie-talkies

Broadcast television

Wireless medical telemetry

Broadcast television

Mobile telepone

Cordless telephone

Pagers

Highway toll tags

Military and federal government

| 400 | 500 | 600 | 700 | 800 | 900 | 1 GHz | 1.1 |

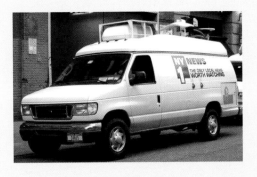

Broadcasting Local News Have you ever wondered how local news correspondents get to the scenes of crimes or fires so quickly? The answer is simple: the assignment desk at a local news station monitors the police and other emergency service radios and sends a news truck out as soon as a story breaks. All trucks have microwave dishes, which send video back to the studio. However, because of the heights of city buildings, the trucks do not always have a clean line of sight to receiving antennas. As a result, they will either bounce the microwave signals off buildings or use a helicopter as a relay.

National version (for viewers outside New York): *ABC's 9.1-meter C-band uplink dish beams the television signal to two satellites. (Using two satellites prevents interruptions to programming in the event that one malfunctions.)*

Local affiliates across the country receive the signal. These affiliates then:
- *distribute an over-the-air signal that can be picked up either by the viewer or by a cable company's head end.*
- *send the signal via cable to the local cable company's head end.*
- *distribute the signal to a DTV head end, which in turn sends the signal up to a DTV satellite.*

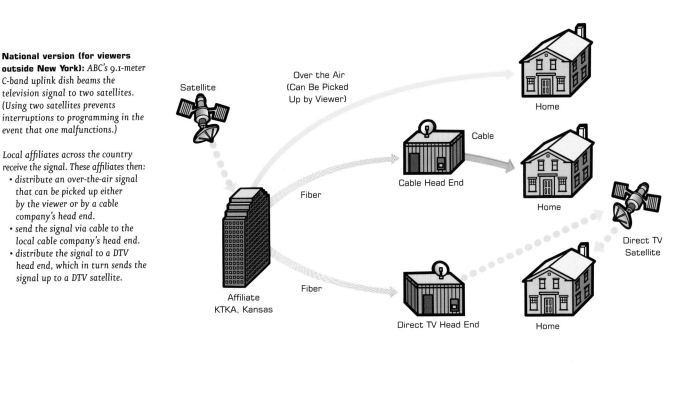

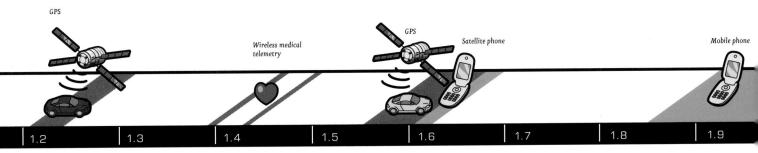

1.2 1.3 1.4 1.5 1.6 1.7 1.8 1.9

The Airwaves

Cable Television In general, broadcast television is well suited to metropolises like New York, both because of the large concentration of people within the traditional 75-mile reception area and because of the higher than average incomes, which mean a healthy market for advertised products. But New York has historically presented a particular challenge for television broadcasters: the large number of very tall buildings within Manhattan negatively impacts reception and in recent years has opened the door to a variety of alternative means of delivering television signals.

Cable television is the most prominent alternative to traditional broadcast technology. Cable companies pick up broadcast signals from terrestrial and other stations, and then amplify and retransmit them through a network of cable that runs under city streets or along utility poles.

Monthly subscriber fees generally cover the cost of both the delivery infrastructure and the programming itself.

Although there are technically nine separate cable television franchises in operation across the five boroughs, only two commercial cable companies hold them: Cablevision and Time Warner. These franchises are granted by the city's Department of Information Technology and Telecommunications (DoITT) and run for a period of 10 years from their inception in 1998. They provide the franchisee with the right to serve a clearly defined geographic area; in return, these companies must commit to serve not only 100 percent of the residences within their areas but also community destinations such as firehouses, senior centers, and schools. The city receives roughly 5 percent of the gross revenue earned in the area as a franchise fee.

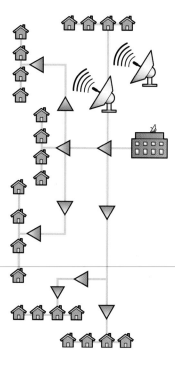

How cable works *Cable systems receive programming via terrestrial or satellite receiving antennas, or may mount their own community-based programming. From here, often referred to as the head end, the trunk part of the system transports the signal to the neighborhood, either through coaxial cable or, increasingly, via fiber. From there, a distribution cable in the neighborhood runs past the homes of subscribers and is connected to flexible drop cable that is routed to subscribers' residences. Specialized amplifiers are located as necessary to increase the signal level (without increasing distortion) for delivery to multiple homes.*

A Snapshot of Cable Franchises

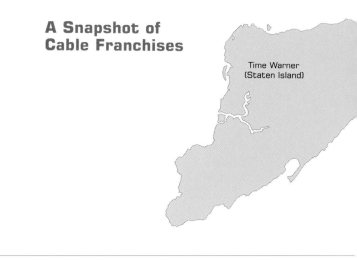

Time Warner (Staten Island)

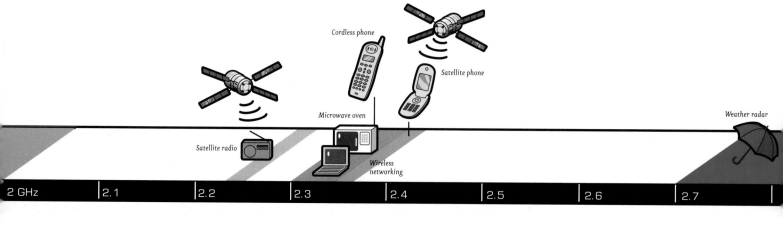

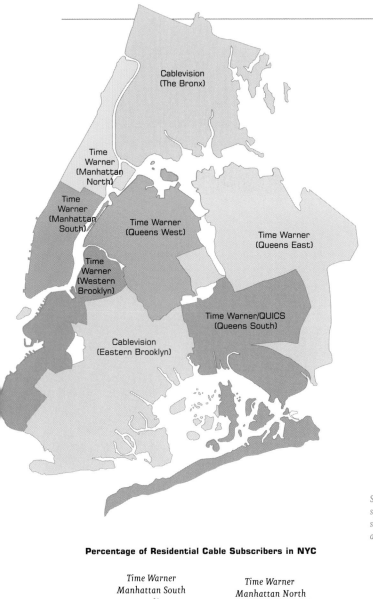

Percentage of Residential Cable Subscribers in NYC

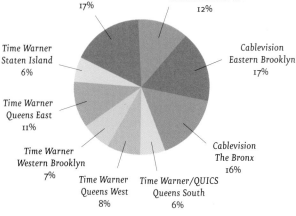

Satellite Broadcasting

Although satellite television is less predominant in New York than in other urban markets across the country, the use of satellites remains a critical part of the region's telecom infrastructure. Without satellites, both the extent of programming for broadcast and cable television and the quality of it would deteriorate severely.

Satellite technology, in its simplest form, involves the transmission of broadcast signals from satellites orbiting the earth. These satellites are all in what is referred to as "geosynchronous orbit": they revolve around the planet at the same speed as the earth turns, effectively staying in one place relative to the earth's surface. Satellite dishes attached to commercial buildings or homes are directed at the satellite when installed, and from that point on—at least theoretically—are able to pick up an uninterrupted signal relayed from other parts of the globe.

Satellite Basics

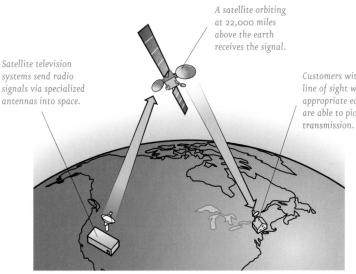

The Airwaves

Radio Technology Radio broadcasting in New York is technically no different than it is in any other city. Sound waves disturb an electrical current in a microphone and travel to a transmitter, which transmits the resulting radio signal to an antenna. From there, the signal is sent out into the air as radio waves, accessible to anyone in the area with a receiver tuned to that channel.

As elsewhere, there are two types of common radio transmissions in New York—AM and FM. AM transmission operates at lower frequencies (540 to 1700 kHz) and can therefore travel much greater distances. It is carried both via ground waves and sky waves, which are reflected back to earth. The picture, or video, part of the television signal as well as many radio stations broadcast via AM transmission.

In contrast, FM transmissions operate at higher frequencies (88 to 108 MHz). Because they are not reflected by the earth's atmosphere, they tend to travel shorter distances—generally from 15 to 65 miles. But they are also less prone to static or interference and commonly produce a better reproduction of the originally broadcast sound. The audio part of a television signal, many radio stations, and nearly all of the wireless appliances found in New York rely on parts of the FM spectrum.

Amplitude Modulation (AM)

Frequency Modulation (FM)

AM vs. FM Most New Yorkers are aware that there are two different ways of transmitting radio signals—AM and FM. Amplitude modulation (AM) involves combining the carrier wave and electric wave in a way that permits the amplitude, or strength, of the carrier wave to match the change in the electric wave. In contrast, frequency modulation (FM) involves combining the carrier and electric waves so that the frequency of the carrier wave changes to match that of the electric wave.

Fixed satellite communications

Radio Station Locations Although the metropolitan region's radio stations are broadcast from a variety of locations, the greatest concentration of them remains in midtown Manhattan, at the Empire State Building.

Station Wattage

o   10–4,000 watts
O   4,001–10,000 watts
◯   10,001 + watts

●   *FM Stations*
●   *AM Stations*

WHUD
WKHL
WXPK
WFAS
WVNJ   WRTN
WVOX
WWRV
WWDJ
WMCA   WFDU   WZRC
WADO   WBBR   WHCR   WFAN
WABC   WPAT   WWRU   WJUX   WCBS
WDHA   WKDM
WFUV
WOR   WSNR   WEPN   WNYU   WCWP
WINS   WLIB   WKCR
WHPC
WFME   WFMU   WNYC   WHPC
WKTU
WSOU   WWRL   WQEU
WQCD   WTHE
WBGO   WHTZ   WKJY
WNSW   WHLI   WRHU
WJDM   WNYE
WGBB
WSIA
WKRB

There has been an antenna on top of the Empire State Building since its opening in 1931. NBC sent its first experimental television signal from the top in December of that year. Ten years later, it began broadcasting the first commercial television station from the building.

The building's dirigible mast, now the base of the television tower, was originally designed as a mooring mast for air blimps. Insufficient consideration was given to the volatile wind conditions at that altitude, however, and after several unsuccessful attempts at mooring blimps, the idea was abandoned. The 22-story, 222-feet, 60-ton mastlike structure was constructed on top of the building in 1950.

The building was the site of the first master FM antenna system in the world, designed to allow multiple FM stations to broadcast simultaneously from one source. Signals are transmitted in a circular pattern that covers the region and extends to areas some 65 miles away. Today, a number of television and radio stations continue to broadcast from the building's antenna, including CBS, ABC, NBC, PBS, Fox, and UPN. In addition to its role in television and radio broadcasting, the Empire State Building houses telephone transmission facilities and serves as the focal point for network television remote relays, paging services, microwave operations, and closed-circuit broadcasts of movies and sporting events.

Radio Stations Broadcasting from the Empire State Building

- WAXQ
- WBAI
- WBLS
- WCAA
- WCBS
- WHTZ
- WLTW
- WNEW
- WNYC
- WPAT
- WPLJ
- WQCD
- WQHT
- WQXR
- WRKS
- WSKQ
- WWPR
- WXRK

Transmission The physical transmission of radio signals, including broadcast television signals, occurs across the metropolitan region 24 hours a day. Prior to 9/11, most of the broadcast networks sent television signals via the antenna at the World Trade Center. Since then, the networks have generally relied on the Empire State Building, long home to FM radio stations. Signals sent from the building are line-of-sight transmissions, which means they are limited to a 60–70 mile range. (Lower-powered FM signals sent out locally from college stations in the city can reach anywhere from a mile to 20 or 30 miles in some instances.)

Most of New York's AM stations have transmitters in New Jersey, rather than New York; exceptions are WCBS and WFAN, which transmit off City Island, and WQEW in Maspeth, Queens. AM transmission, unlike FM, is not dependent on line of sight: AM waves generally travel along the ground during the day, and reception is a function of ground connectivity. Because salt water is a very good conductor, both WCBS and WFAN, transmitting from City Island, can be heard all the way up Long Island Sound, out onto Cape Cod and as far away as Bermuda.

Transmission at night is another story. A charged layer of the atmosphere can bounce signals hundreds or even thousands of miles. Because of the potential for interference due to this extended reach, regional stations often reduce transmitter power between sunset and sunrise. At one time, the FCC had selected one "dominant" station to operate on "clear channel" frequencies at night in order to provide clear service to a wide area—roughly 750 miles in radius. While the prohibition on other AM stations in the area broadcasting at night has largely been lifted, those choosing to broadcast at night often use directional antennas to avoid conflicting with other stations.

Even during the day, transmission must often be manipulated to avoid interference in neighboring markets. Most AM stations commonly use directional antennas that limit their signals in certain directions and enhance them in others. For example, WEPN 1050 uses three towers at its site near Giants Stadium in New Jersey to direct most of its signal over Manhattan and Long Island and to minimize signal oriented toward Philadelphia.

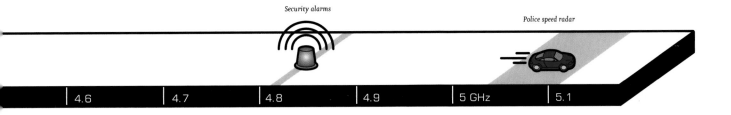

Security alarms

Police speed radar

4.6   4.7   4.8   4.9   5 GHz   5.1

KEEPING IT CLEAN

Two hundred years ago, New Yorkers struggled to fight off cholera and yellow fever; today they work at keeping even the smallest of fish alive in the Hudson River. New York is cleaner than ever, and with 10 million people living in proximity to one another, that is no small feat. Just how three century-old systems are able to keep it clean is perhaps the greatest of New York's infrastructure secrets.

The water system, a marvel of nineteenth-century engineering, delivers over a billion gallons of water each day to city consumers. The companion sewer system, over 6,000 miles of underground pipes and treatment plants, carries that water away once used. These are complemented by a garbage system that lives up to the task of finding a resting place—out of the city—for every one of the 25,000 tons of garbage New Yorkers produce each day.

New York City's water
is renowned—for its
abundance and its purity.
Both are products of

Water

the intricate system of dams, reservoirs, tunnels, and aqueducts that carry 1.3 billion gallons of water each day from protected watershed areas in upstate New York to residents of the five boroughs. The system also supplies, en route, another 100 million gallons of water daily to over a million customers in neighboring counties.

New York's water wasn't always so clean or so plentiful. In the early seventeenth century, water to meet the needs of the small town at the tip of Manhattan was drawn from what was known as "the Collect Pond," a 48-acre spring-fed pond located near Franklin and Pearl Sts. in lower Manhattan. But the absence of any organized system of disposing of human and animal waste meant that locally drawn well water was often contaminated.

The health impacts of unsanitary water were recognized as far back as 1799, when a charter was granted to a private company —the Manhattan Company—to provide freshwater to paying customers from wooden pipes laid just beneath the surface of the streets in lower Manhattan. But even privately supplied water was ultimately bad tasting and often impure—and did nothing to stem the tide of yellow fever, which hit the city in 1819 and 1822, or of cholera, which ravaged the city in 1832 and again in 1834.

Clean water arrived in New York in 1842, with the opening of the Croton Aqueduct system. The Croton River in Westchester County was dammed, and the reservoirs created fed an underground aqueduct that tunneled 30 miles southward—first to the Yorkville receiving reservoir (now Central Park's Great Lawn) and then on to the Murray Hill distributing reservoir (now the site of the New York Public Library) at West 42nd St. Entirely gravity-fed, the system could provide 30 million gallons a day— enough to meet the growing city's needs through the turn of the century.

But the introduction of sewers, flush toilets, and household faucets in the late

Completed in 1842, the Murray Hill distributing reservoir was located at 42nd St. and Fifth Ave., on the current site of the New York Public Library. Built in the Egyptian Revival style, its 45-foot-high granite walls and two holding tanks had a capacity of 24 million gallons of water.

nineteenth and early twentieth centuries caused a dramatic jump in water use, well beyond the level that could be provided by the Croton system. To meet this need, a state law was passed in 1905 allowing the city to purchase land in the Catskills. Residents of several villages were relocated and their towns flooded to create the Ashokan Reservoir. From the reservoir, a 92-mile-long aqueduct was blasted through rock, and under the Hudson River, to bring the water to the edge of the city.

The Origins of Chase Bank

The first organized delivery of water to the municipality was a function of private enterprise. In 1799, the New York State legislature gave Aaron Burr's newly formed Manhattan Company the exclusive right to supply water to the city—initially via wooden pipes. But rather than bring water from outside as planned, the company sank more wells locally and stored it in a reservoir at Chambers St.; thus the quality of its water was no better than that drawn directly from the Collect Pond itself. The company prospered nevertheless and used its surplus to start a bank—the Bank of the Manhattan Company—that was more profitable than its water delivery business. As its banking operations expanded, its water delivery operations shrank, and in 1808 the company sold its water operations to the city. The Bank of the Manhattan Company would see its way through a variety of new alliances, most recently the merger of Chase Manhattan and J. P. Morgan.

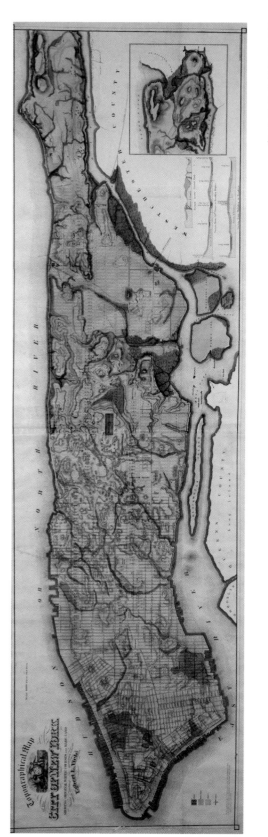

Mapping Manhattan's Waters
Egbert Ludovicus Viele's water map, published in 1889 and still used today by engineers and developers, mapped out all of the city's original waterways in great detail.

Water

New York's Water Delivery Network

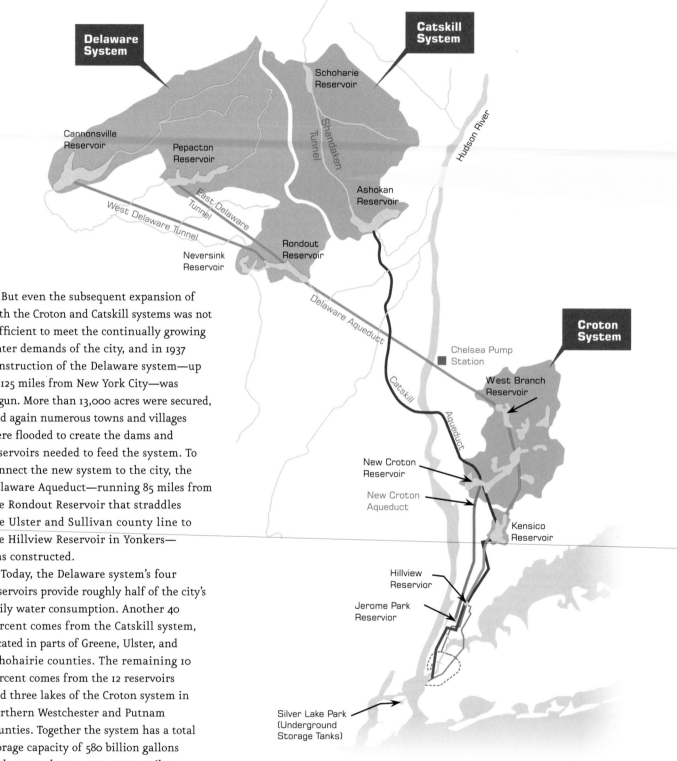

Delaware System

Catskill System

Schoharie Reservoir

Shandaken Tunnel

Hudson River

Cannonsville Reservoir

Pepacton Reservoir

Ashokan Reservoir

West Delaware Tunnel

East Delaware Tunnel

Neversink Reservoir

Rondout Reservoir

Delaware Aqueduct

Chelsea Pump Station

Croton System

West Branch Reservoir

Catskill Aqueduct

New Croton Reservoir

New Croton Aqueduct

Kensico Reservoir

Hillview Reservoir

Jerome Park Reservior

Silver Lake Park (Underground Storage Tanks)

But even the subsequent expansion of both the Croton and Catskill systems was not sufficient to meet the continually growing water demands of the city, and in 1937 construction of the Delaware system—up to 125 miles from New York City—was begun. More than 13,000 acres were secured, and again numerous towns and villages were flooded to create the dams and reservoirs needed to feed the system. To connect the new system to the city, the Delaware Aqueduct—running 85 miles from the Rondout Reservoir that straddles the Ulster and Sullivan county line to the Hillview Reservoir in Yonkers—was constructed.

Today, the Delaware system's four reservoirs provide roughly half of the city's daily water consumption. Another 40 percent comes from the Catskill system, located in parts of Greene, Ulster, and Schohairie counties. The remaining 10 percent comes from the 12 reservoirs and three lakes of the Croton system in northern Westchester and Putnam counties. Together the system has a total storage capacity of 580 billion gallons and covers about 2,000 square miles—roughly the size of the state of Delaware.

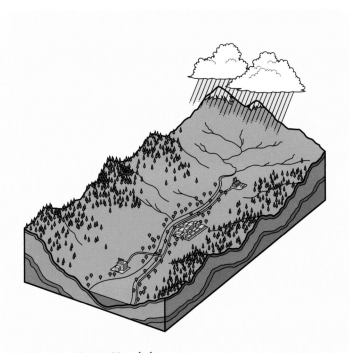

Watershed Basics *Watersheds are generally defined as areas of land that drain into the same water course—generally rivers, lakes, or underground aquifers. Often they cross county or state boundaries.*

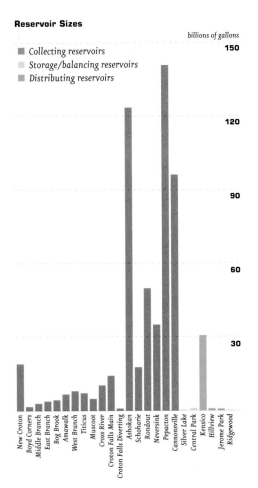

Reservoir Sizes

billions of gallons

- ■ *Collecting reservoirs*
- ▨ *Storage/balancing reservoirs*
- ■ *Distributing reservoirs*

New Croton · Boyd Corners · Middle Branch · East Branch · Bog Brook · Amawalk · West Branch · Titicus · Muscoot · Cross River · Croton Falls Main · Croton Falls Diverting · Ashokan · Schoharie · Rondout · Neversink · Pepacton · Cannonsville · Silver Lake · Central Park · Kensico · Hillview · Jerome Park · Ridgewood

| Reservoirs |

The New York City water system relies on 18 collecting reservoirs as well as two storage reservoirs and four distributing reservoirs, the latter within the city's boundaries. The collecting reservoirs are spread across the three watersheds: Croton, the smallest of the three systems, has 12 reservoirs and three lakes; the Catskill has two sizable reservoirs; and the Delaware system encompasses four. The individual capacity of the reservoirs varies dramatically, with the smaller reservoirs in the Croton system holding anywhere from three to 10 billion gallons and the largest—the Pepacton Reservoir in the Delaware system—holding 140 billion gallons.

The creation of these reservoirs during the nineteenth and first half of the 20th centuries involved the partial flooding of some 30 separate communities in Sullivan, Delaware, Ulster, and Putnam counties. Over 9,000 people were displaced in the process and an estimated 11,500 graves dug up and reinterred. But the politics went beyond just the decimation of homes and graveyards: the State of New Jersey appealed to the Supreme Court in the late 1920s —unsuccessfully—to enjoin New York City and State from using the waters of any Delaware River tributary, even those physically located within New York State.

Water

Aqueducts To move water from the three upstate watersheds to city users, four spectacular aqueducts were constructed. The earliest, the Old Croton Aqueduct, was abandoned in 1955, and much of its path is now used by bikers and walkers. The three others—the New Croton, Catskill, and Delaware aqueducts —are in operation today and remain among the most ambitious of New York City's engineering achievements.

- **The New Croton Aqueduct**, running from the original Croton Reservoir system south to Jerome Park Reservoir in the Bronx, was constructed between 1885 and 1890, and is notable for including "weepers"— four inch by eight inch openings in the brick tunnel wall, through which groundwater can drain into the tunnel if desired.

- **The Catskill Aqueduct**, reaching from the Ashokan Reservoir in the Catskills to the Kensico receiving Reservoir in Westchester County, features an extraordinary river crossing—an 1,100-foot-deep water tunnel through bedrock between Storm King Mountain on the west side of the Hudson and Breakneck Ridge, near Beacon, on the east side.

- **The Delaware Aqueduct**, the newest of the city's aqueducts, has the distinction of being the world's longest continuous underground tunnel. Thirteen and a half feet wide, it runs 85 miles from the Rondout Reservoir, which straddles the Ulster/Sullivan County line in upstate New York, to Hillview Reservoir in Yonkers.

A Trio of Aqueducts
To contend with various geological conditions, sections of the aqueducts serving New York City were built at different depths along the route toward the city.

Aqueduct Shapes
The three aqueduct systems serving New York are made up of sections of varying shapes, some of which are shown here in cross section.

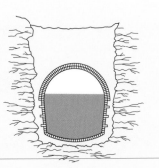

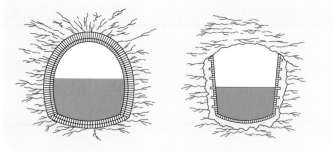

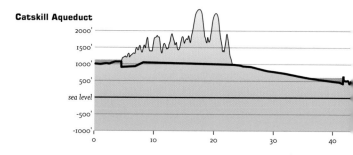

Catskill Aqueduct

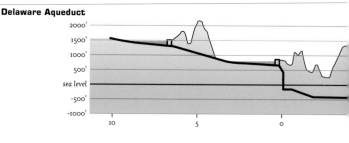

Delaware Aqueduct

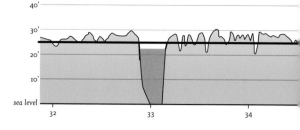

Croton Aqueduct

Water Under the Hudson

The Catskill tunnel crosses the river at one of its deepest points—between Storm King, south of Newburgh, and Breakneck Ridge, just south of Beacon. Identifying the appropriate depth at which to cross the river was a complex task.

The shafts of the tunnel were sealed with forged-steel caps weighing 46 tons each.

To identify the location of the bedrock, test shafts were sunk on each side of the river. From a depth of 300 feet down each shaft, drills worked toward each other at a 43-degree angle, meeting at 1,500 feet. When a similar exercise was successfully repeated at a shallower angle, engineers were convinced that it was safe to locate the tunnel at a depth between the two meeting points.

A 17-foot diameter circular tunnel was drilled between the two shafts at a depth of 1,100 feet.

To ensure stability, the tunnel needed to be placed under at least 150 feet of granite.

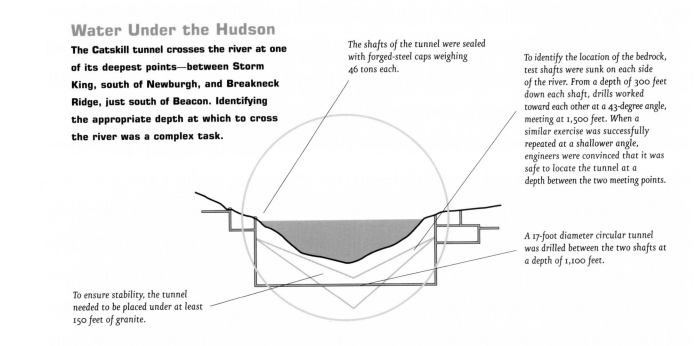

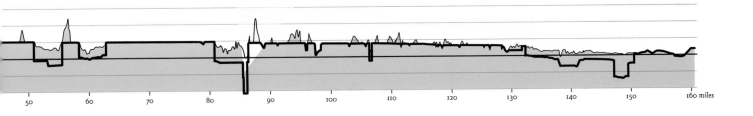

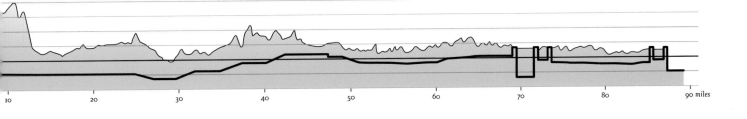

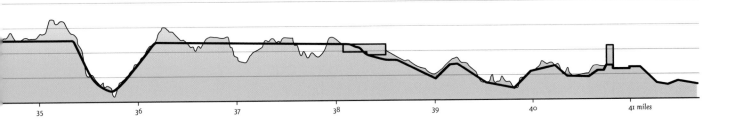

Water

Local Distribution

Distribution of water within the city relies on two large tunnels that crisscross the five boroughs: City Water Tunnel No. 1 and City Water Tunnel No. 2. City Water Tunnel No. 1 was completed in 1917 and stretches 18 miles from the Hillview Reservoir in Yonkers through the Bronx and across the Harlem River into Manhattan. It travels under Central Park and continues under the Lower East Side, crossing beneath the East River to Brooklyn and terminating at Third Ave. and Schermerhorn St. in Brooklyn. Covered in most places by at least 150 feet of rock, it ranges from 200 to 300 feet deep and carries between 500 and 600 million gallons of water each day.

City Water Tunnel No. 2 was constructed during development of the Delaware Aqueduct system and went into use in 1936. With a capacity of between 700 and 800 million gallons per day, it runs from the Hillview Reservoir in Yonkers south through the Bronx, crossing the East River to Queens near Astoria. From there it travels southwest, connecting up with the five-mile-long Richmond Tunnel, which runs under Upper New York Bay and carries Staten Island's water supply. Like Tunnel No. 1, City Water Tunnel No. 2 has been in continuous use since it was put into service.

Both Water Tunnels No. 1 and No. 2 are in serious need of repair. However, partly due to fears that a reduction in pressure might damage some part of the system, neither can be shut down for maintenance work. To allow these conduits to be taken out of service and to generally add capacity to the system as a whole, City Tunnel No. 3 is under construction currently and will be completed by 2020.

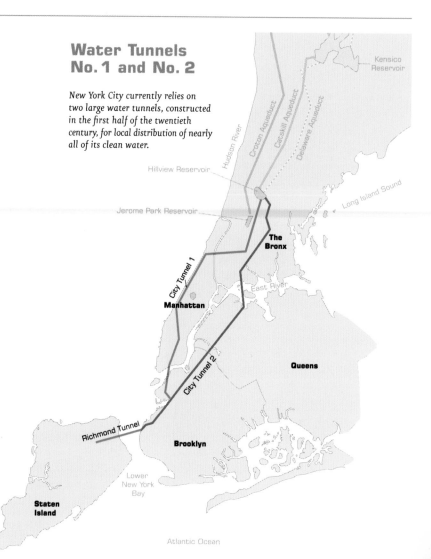

Water Tunnels No. 1 and No. 2

New York City currently relies on two large water tunnels, constructed in the first half of the twentieth century, for local distribution of nearly all of its clean water.

Regulators & Pressure

Like the aqueduct system that carries water to the city's borders, distribution of water to city residents is almost entirely dependent on gravity alone: water reaches the Hillview Reservoir at 295 feet above sea level and from there travels with sufficient pressure to reach the sixth floor of most buildings in the five boroughs. Water travels through the high-pressure mains of the system at a pressure of between 90 and 100 pounds per square inch (psi) and must be supplied to hydrants at a minimum pressure of 35–60 pounds per square inch.

In fact, water pressure in the system is so great that in most cases it must be lowered before it comes to the surface or, like a geyser, it would blow the top off its exit shafts. In the case of Croton water, which is distributed primarily to the lower elevations of Manhattan and the Bronx, pressure (largely determined by the water level of the Jerome Park Reservoir) is naturally reduced by the elevation of, or distance to, the distribution zone being served. However, in the case of Catskill and Delaware water, pressure must be mechanically reduced by regulators at the interconnections between the trunk grid and the distribution network.

Only a small amount of New York City water, between 3 percent and 5 percent, requires additional pressure to reach its intended end users. Pumping stations are located at three elevated areas: Washington Heights, in northern Manhattan, Douglaston, in eastern Queens, and Grimes/Todt Hill on Staten Island.

In-City Water Distribution

Service lines *Branch or service lines connect the city distribution system to the plumbing system within each building and range from one-inch copper tubes to eight-inch cast iron pipes.*

Distribution mains *Distribution mains and submains are generally between six and 20 inches in diameter, and are fed by one or more trunk mains.*

Hydrants *Water is distributed to fire hydrants at a minimum pressure of 40 pounds per square inch at the curb.*

Regulator *Regulators inside water shafts have spring-loaded diaphragms that adjust pressure by rising and falling, depending on water consumption, before distribution to end users.*

Trunk or high-pressure mains *Vertical shafts meet the main water tunnels (Water Tunnels No. 1 and No. 2) at roughly one-mile intervals and carry the water to trunk mains, which are generally between 24 and 84 inches wide and operate under high pressure.*

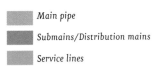

Main pipe

Submains/Distribution mains

Service lines

Water

Hydrants Maintaining water pressure in the city's mains is perhaps most critical to support the network of fire hydrants that form the core of the city's firefighting efforts. There are approximately 118,000 fire hydrants located throughout the city, most of them constructed in like fashion: a branch line from a street main to the sidewalk is attached to a "hydrant bend," and the hydrant appliance sits atop the bend. To access the water, a protective cap is removed, a hose attached, and a valve nut rotated to open the main valve, allowing water from the municipal system to fill up the hydrant's barrel.

Today's fire hydrants are direct descendants of the earliest hydrants, which were sited in lower Manhattan in the early years of the nineteenth century (the first was located on the corner of William and Liberty Sts. in 1808). Innovations since that time have been modest: a high-pressure system—involving fatter hydrants, larger diameter hose nozzles, and higher water pressure—was developed in the early part of the twentieth century to serve taller buildings in Manhattan and Brooklyn, as well as the Coney Island amusement park, but it was largely abandoned by 1980 (although some of the fatter hydrants can still be found around the city). More recently, "Series S" hydrants have been installed. These have a breakaway flange, which helps avoid damage to the water main pipe when a hydrant is struck by a motor vehicle.

Most fire hydrant systems require a minimum pressure in the neighborhood of 35–60 psi, to ensure functionality. In New York, the absolute minimum is 40 psi, although the municipal water delivery system as a whole is designed to ensure higher levels of between 45 and 60 psi.

The Modern Hydrant

A custodial lock, a muffin-shaped cap on top of the hydrant, covers the top of the main valve and closes off the water main feeding the hydrant. The new locks have a raised "x" on top.

The new hydrant locks can be opened only by firefighters. Tens of thousands have been installed to prevent tampering. Special spray caps, which limit water flow from 1,000 gallons a minute to 25, can still be secured from the Fire Department by applying in writing at local engine companies.

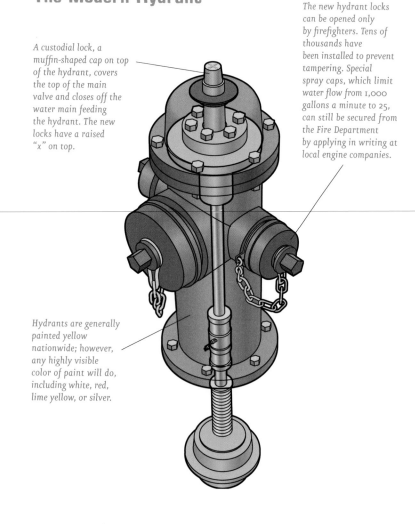

Hydrants are generally painted yellow nationwide; however, any highly visible color of paint will do, including white, red, lime yellow, or silver.

Plug Uglies

In the first half of the nineteenth century, something of the frontier spirit existed in New York with respect to firefighting. Hydrant companies were formed to manage newly installed fire hydrants, and they rapidly entered into shifting alliances with engine companies and hose companies—whose equipment was necessary to actually put out a fire. Getting to the hydrant first, during a fire incident, was key to a hydrant company's success. To boost success rates, hydrants were occasionally covered by barrels to disguise their locations from competing companies. Alternatively, hydrants would be protected by gangs, known affectionately as "plug uglies."

Water Storage

Despite the enormous capacity of the city's upstate network of reservoirs, having enough water is a constant issue for the Department of Environmental Protection. Droughts occurred as far back as 1952, regularly during the 1960s, and again on three separate occasions during the 1980s. In each case, a variety of measures were taken: in 1952, the city attempted to seed clouds above the Catskill reservoirs; in 1981, New Jersey water was delivered to the city by a large pipeline cabled to the lower deck of the George Washington Bridge. And at least twice in recent history, in 1985 and again in 1989, the city used its Chelsea pumping station, located 65 miles north of the city near Poughkeepsie, to draw and filter Hudson River water for municipal consumption.

To ensure it meets the needs of residents even during the least rainy periods of the summer, a number of initiatives are under way. For some time, the department has been evaluating drawing undrinkable groundwater from wells in Brooklyn and Queens to supply water at a discounted rate to businesses within the city that need water for manufacturing purposes. It has also had discussions with big commercial users of water about switching, where possible, to recycled wastewater for nondrinking purposes; the Port Authority, for example, uses tens of thousands of gallons of water each day to wash planes at both LaGuardia and Kennedy airports.

In emergencies, the city has the ability to draw from river water and also to draw from neighboring water systems. The Chelsea pumping station near Poughkeepsie can filter and funnel roughly 300 million gallons of water a day to the West Branch Reservoir near Carmel. In addition, the city's supply lines are interconnected with those of the lower Delaware system serving Philadelphia and central New Jersey. If downstream users in these areas had sufficient alternative supply, the city could draw an additional 300 million gallons a day from this system.

Silver Lake What some might term the "belt and braces" approach is used to ensure that Staten Island has enough water. Connected by the five-mile-long, 10-foot-wide Richmond Tunnel to City Water Tunnel No. 2 in Brooklyn, the borough can and does—like other parts of the city—draw directly on the water supply from Hillview Reservoir. But it also has a stash of water for emergencies: two 50-million-gallon tanks—known as the Silver Lake storage tanks—keep a two-day supply on hand 20 feet beneath the island. The tanks, considered the world's largest underground storage tanks, replaced the Silver Lake Reservoir, which attracted too many birds and consequently too many bird droppings.

History of Drought and Water Consumption

million gallons per day no restrictions watch warning emergency

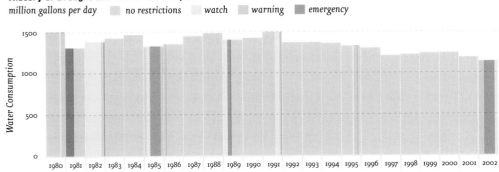

Water

 It is not entirely surprising that a water system as extensive, and as old, as New York City's should spring a leak now and again. This is particularly true with respect to the in-city delivery system and its trunk mains, where high pressure puts a constant strain on the pipes involved. Roughly half of the distribution system was built prior to 1930 and thus consists of unlined cast iron pipes; the remaining half, built subsequently, is stronger due either to new forms of iron or concrete linings that greatly reduce susceptibility to corrosion and failure.

While water main breaks within the city regularly garner the attention of the local press, they rarely endanger the integrity of the overall water system. More important to the system as a whole, though less newsworthy, are the substantial cracks in the aqueducts feeding the city's reservoirs. Together, these cracks leak an estimated 36 million gallons of water each day—a pittance compared to the 1.3 billion gallons successfully consumed, but a serious concern to engineers worried about the structural integrity of the rock and soil through which these tunnels have been burrowed.

The subject of greatest concern to city engineers is the tunnel that provides most of the city's water: the Delaware Aqueduct. Running at depths from 300 to 2,400 feet below the surface, the aqueduct has sprung confirmed leaks in several places—including one on the northern edge of Newburgh, near where the Aqueduct crosses the Hudson. Close to the town of Roseton, a leak has created a freshwater spring, a four-foot deep lake, and a 35-foot sinkhole. This is of significant concern to water system engineers, as unstable rock formations—noted nearby during the underwater tunnel's construction in the early 1940s and again during the last dry inspection, in 1958—could lead to collapse of part of the aqueduct.

Dry inspections of the aqueduct are no longer possible: engineers fear that without the supporting pressure of the water from inside, a wall weakened by water leakage might crumble. Instead, to better understand and map the leaks in the Delaware Aqueduct, tunnel inspections have been carried out by robotlike devices that travel through the aqueduct and record the condition of various segments of the tunnel.

A Profile of New York's Water Mains

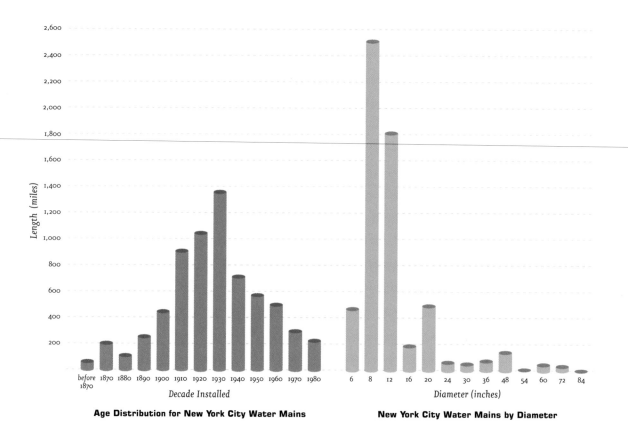

Age Distribution for New York City Water Mains   New York City Water Mains by Diameter

In Search of a Leak To better understand the nature of leaks emanating from the Delaware Aqueduct, a tunnel inspection company teamed up with Woods Hole Oceanographic Institute on Cape Cod to design a specialized torpedo-shaped robot that could maneuver through the aqueduct's pipes and record their condition. In June 2003, the robot successfully completed a 15-hour, 45-mile journey through the Rondout–West Branch Tunnel of the aqueduct. Data gathered on that journey are intended to be used to both map areas of leakage and develop a repair plan.

The submarine's computer measured changes in currents, sounds, the earth's magnetic field, and the tunnel's water pressure, and took photos.

Whiskerlike titanium wires protruded from the robot's nose, to soften any collisions with tunnel walls.

The craft was designed and programmed, using navigational aids and acoustic beacons, to stay dead center in the tunnel, so that five cameras in its nose could have a 360-degree view of its walls.

The torpedo-shaped craft was nine feet long and 16 inches in diameter, and weighed 800 pounds.

Water

Water Tunnel No. 3 Perhaps no project has captured the fancy of New Yorkers like the construction of Water Tunnel No. 3, which has been under way for the last 35 years and will not be completed for at least another 15. The largest construction project in the city's history, and among the most complex engineering projects under way in the world today, completion of the 60-mile tunnel is expected to cost a total of six billion dollars.

Planning for the tunnel began in the 1950s, when it was recognized that a third tunnel was needed to reduce reliance on the existing tunnels and allow them to be taken off-line for necessary repair and maintenance work. Phase 1 of the project—which involved the construction of three huge valve chambers, 14 supply shafts, and a tunnel running south from Hillview Reservoir through the Bronx, Manhattan, Roosevelt Island, and into Astoria— was begun in 1970, largely completed in the 1980s, and put into service in 1998. Unlike the current situation, where control valves are at tunnel level and inaccessible, valves for the new tunnel are housed in highly accessible underground chambers—the largest of which, the Van Cortlandt Park Valve Chamber, is 620 feet long (longer than two football fields laid end to end).

The second phase of the project is under way currently. It involves two segments: that in Manhattan, which will run from Central Park south along the West Side and then east to the South St. Seaport and north to East 34th St.; and a second in Brooklyn/Queens, which will run through Red Hook, Maspeth, Woodside, and Astoria and will connect with the Richmond Tunnel serving Staten Island.

The final phases of the project, Stages 3 and 4, are still in the planning and design phase. Stage 3, in contrast to the first two stages, will actually bring new capacity to the system. It involves the construction of a new aqueduct connecting the Kensico Reservoir in Westchester County with the Van Cortlandt Park Valve Chamber, which will provide an alternative route for water from the Delaware and Catskill aqueducts to reach the city and will operate at greater pressure than the current line due to the higher elevation of Kensico. Stage 4 involves the construction of a new 14-mile-long stretch from the new valve chamber under Van Cortlandt Park southeast through the Bronx and over the East River into Flushing.

Man and Machine The tunnel-boring machine hard at work under Manhattan's bedrock is very similar to the machine that was used to dig the "Chunnel," connecting the south coast of England and France. Known as "the mole," it relies on a series of steel-cutting tools that continuously rotate to chip off sections of rock to achieve forward movement of about 50 feet each day— more than twice the rate achieved during Stage 1, when conventional drilling and blasting techniques were employed. So large is the machine— its body alone is 50 feet long— that it had to be lowered through tunnel shafts in sections and assembled at the bottom.

As interesting as the machine itself are the men who have mastered its operation. Known as "sandhogs," they carry out some of the most dangerous work anywhere in city government: 24 of them died during construction of the first phase of Water Tunnel No. 3. Among their number are numerous descendants of the newly arrived immigrants who carved the Catskill Aqueduct, which runs deep under the Hudson River, and of those who dug the Holland Tunnel, the region's first vehicular tunnel. Though the work remains dangerous and dirty, the pay has improved; today's sandhogs earn up to $120,000 a year.

Valve Chamber

Valve chambers allow engineers to manage the flow of water as it moves through the system. The largest valve chamber in the system is the Van Cortlandt Park Valve Chamber, located directly underneath Van Cortlandt Park in the Bronx. The chamber itself is huge—620 feet long, 42 feet wide, and 41 feet high. Within the chamber sit large conduit pipes (eight feet in diameter) with flow meters that will measure, direct, and control the flow of water coming in from the Delaware and Catskill systems.

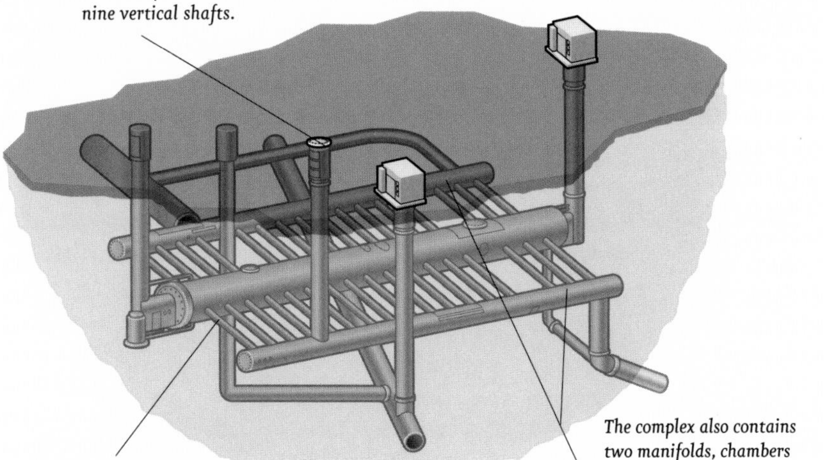

The Van Cortlant Park complex contains nine vertical shafts.

There are 17 steel-lined lateral tunnels in the valve chamber, each over 100 feet long.

The complex also contains two manifolds, chambers through which water can be distributed, each more than 560 feet long and 24 feet wide.

Water Tunnel No. 3

The second phase of construction of Water Tunnel No. 3 is currently under way on the West Side of Manhattan. Completion of this and the remaining two phases is not expected before 2020.

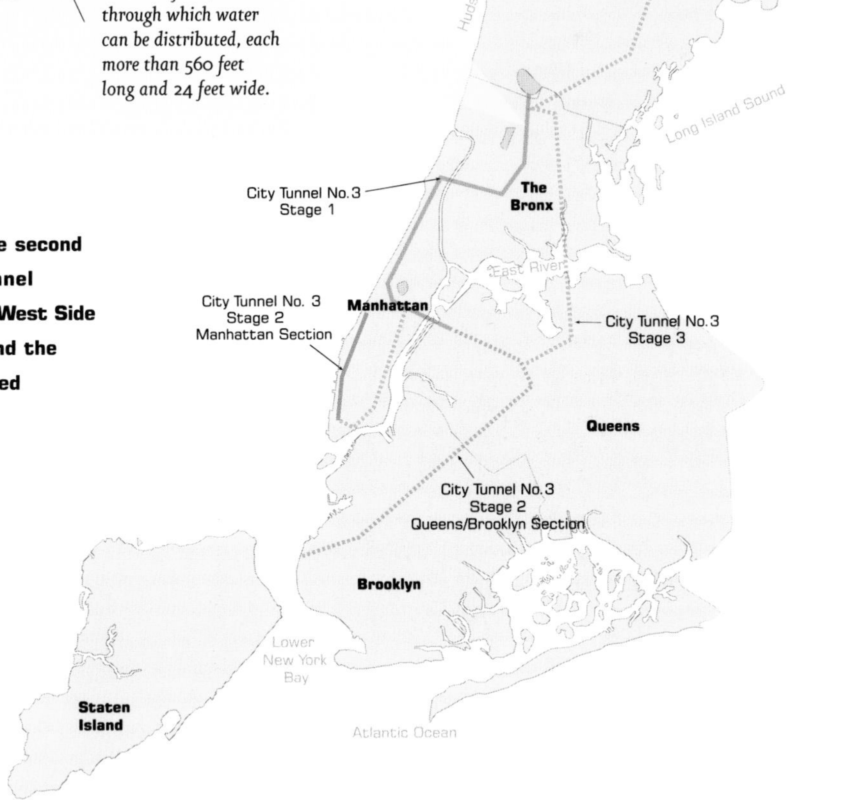

········· Proposed
───── Under construction
········· Completed
───── Activated

Kensico Reservoir

City Tunnel No. 3 — Stage 4

Hudson River

Long Island Sound

City Tunnel No. 3 Stage 1

The Bronx

City Tunnel No. 3 Stage 2 Manhattan Section

Manhattan

East River

City Tunnel No. 3 Stage 3

Queens

City Tunnel No. 3 Stage 2 Queens/Brooklyn Section

Brooklyn

Lower New York Bay

Staten Island

Atlantic Ocean

Water

Water Tanks Getting water to the tops of buildings became a problem in late-nineteenth-century New York, as buildings grew taller and the street mains system—under roughly 50–60 pounds of pressure per square inch—could push it only so far as the sixth floor. The solution to this vertical problem was the rooftop water tower, filled by basement pumps triggered by water depletion. Not only was the tower able to provide water under pressure to even the tallest new buildings, but it also acted as a reservoir, providing spare capacity to handle peaks in demand.

Though New York City is not the only city in the nation to feature water towers, they are more prominent here than almost anywhere else, and the wooden tanks with conical-shaped tops have become a beloved feature of the skyline. An estimated 10,000 to 15,000 of the tanks are currently in use across the five boroughs, a testament to the advanced age of many of the city's tall buildings. Though newer buildings in New York and other cities tend to rely on more modern pump systems, which cycle on and off from the basement when water is called for, rooftop tanks are still considered among the most efficient and reliable means of providing consistent water pressure—particularly in the event of a failed pump.

New York's water towers are generally made of wood and feature cedar planks along the floor and walls. They are easy to construct *in situ*, as no glue or adhesive is needed to hold them together: in barrellike fashion, pressure exerted by galvanized steel hoops surrounding the planks—along with the swelling of the wood that occurs when the tank is filled—prevent any leakage. Wood is a natural insulator, too: three inches of wood have roughly the insulation value of 30 inches of concrete.

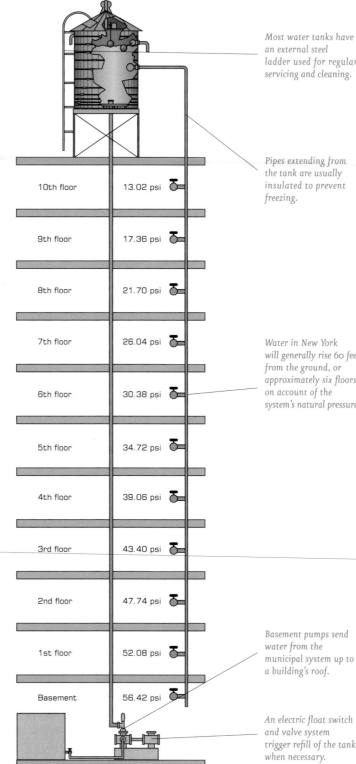

Most water tanks have an external steel ladder used for regular servicing and cleaning.

Pipes extending from the tank are usually insulated to prevent freezing.

Water in New York will generally rise 60 feet from the ground, or approximately six floors, on account of the system's natural pressure.

Basement pumps send water from the municipal system up to a building's roof.

An electric float switch and valve system trigger refill of the tank when necessary.

10th floor	13.02 psi
9th floor	17.36 psi
8th floor	21.70 psi
7th floor	26.04 psi
6th floor	30.38 psi
5th floor	34.72 psi
4th floor	39.06 psi
3rd floor	43.40 psi
2nd floor	47.74 psi
1st floor	52.08 psi
Basement	56.42 psi

Consumption New Yorkers consume plenty of water, though no more or less than those in other cities. In fact, there is some evidence that local per capita water consumption has fallen over the last two decades. At one point in the city's history, in the mid-1980s, close to 1.5 billion gallons were being consumed each day; today, the figure stands at roughly 1.1 billion gallons daily.

A number of factors, beyond just conservation awareness efforts, have contributed to this trend. In contrast to the previous system of flat charges for buildings based on the size of their frontage, use-based pricing—along with meters to implement the program—was introduced across the five boroughs in 1986 to discourage wasteful consumption. In the early 1990s, low-flow toilets, showerheads, and faucets were introduced in a variety of new city housing projects. Simultaneously, a leak-detection program was introduced in tens of thousands of apartments and homes.

Washing machines *use 56 gallons to complete an average load.*

A standard toilet *uses six gallons per flush.*

A dishwasher *will use 24 gallons per load.*

A typical shower *uses seven gallons of water every minute it is running.*

A lawn sprinkler *covering one-fifth of an acre will use 24 gallons each month.*

A faucet *with a slow drip uses 17 gallons of water per day.*

Water by the Foot

All water meters used by the city's Department of Environmental Protection to monitor water usage have odometerlike readings. Though the meters generally read in cubic feet, with one cubic foot of water equivalent to 7.48 gallons, the department charges based on 100-cubic-foot units (HCF): $1.60 per 100 cubic feet provided. (One HCF, therefore, is equivalent to 748 gallons.) Compound meters, designed for buildings with wide variations in consumption, consist of two meters, with an internal control mechanism which diverts water flow to one meter or the other depending on its volume. Each meter has a unique serial number for identification purposes.

Water

Though New York's water is considered pure, thanks to its rural origins, it is by no means untreated. A variety of chemicals are added along the way to ensure a consistent quality upon arrival: aluminum and other chemicals are added to form "floc"—sticky particles that attract dirt and sink to the bottom during sedimentation; chlorine is added in a variety of places en route to kill bacteria; caustic soda and phosphoric acid are added to lower the acidity of the water, making it less corrosive and less likely to interact with lead and copper in building pipes; and fluoride is added, at a concentration of roughly one part per million, for dental hygiene.

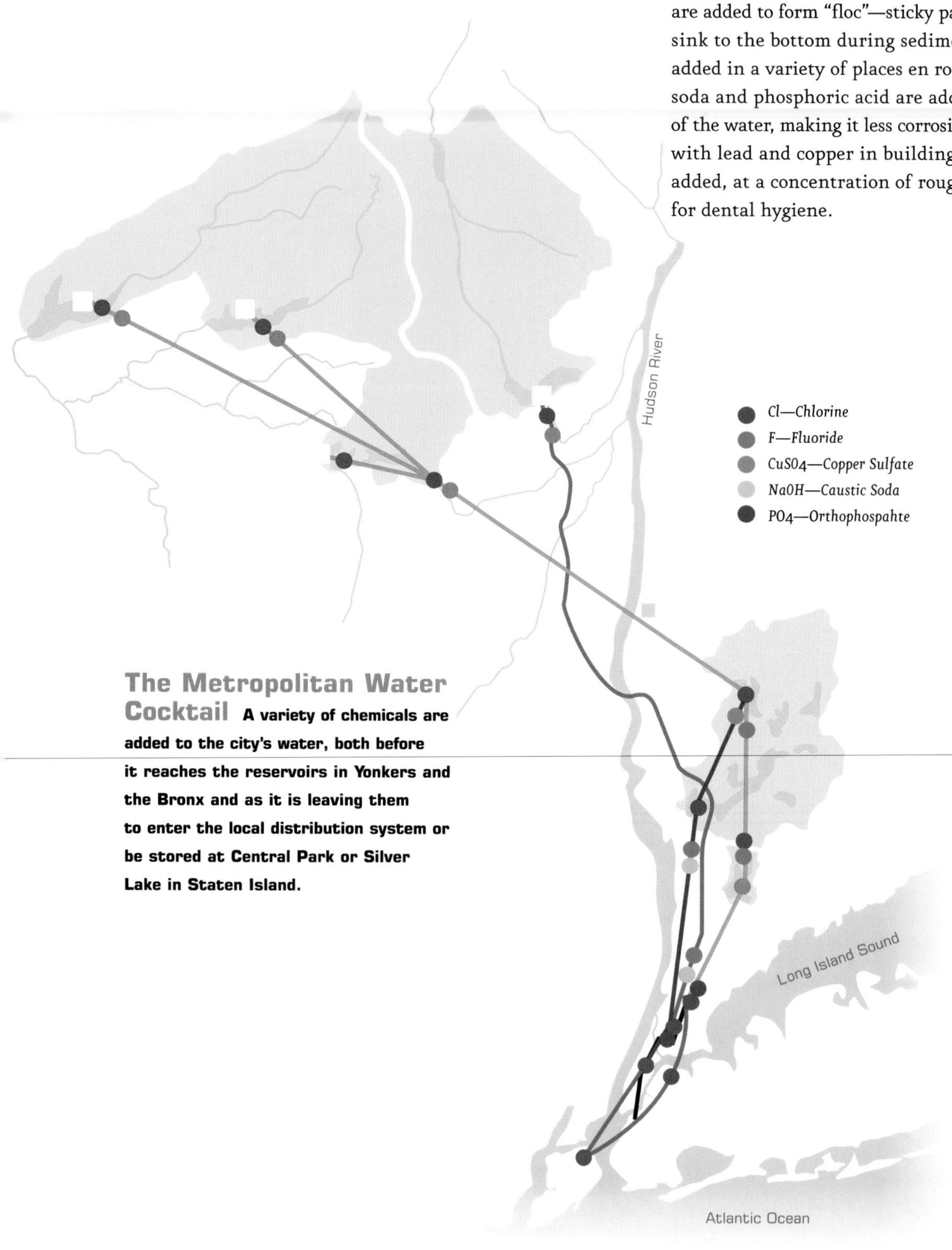

Hudson River

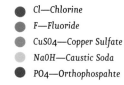

● Cl—Chlorine
● F—Fluoride
● CuSO4—Copper Sulfate
● NaOH—Caustic Soda
● PO4—Orthophospahte

The Metropolitan Water Cocktail A variety of chemicals are added to the city's water, both before it reaches the reservoirs in Yonkers and the Bronx and as it is leaving them to enter the local distribution system or be stored at Central Park or Silver Lake in Staten Island.

Long Island Sound

Atlantic Ocean

To ensure that the appropriate levels of chemicals are being added, and to comply with federal and state drinking water regulations, the city's Department of Environmental Protection regularly samples water quality throughout the city. Two types of sampling sites exist: *compliance* sites, consisting of three stations (that from which the sample is taken and adjacent upstream and downstream stations, which are tested if a positive result is found), and *surveillance* sites, which are located on trunk mains and serve primarily as early warning systems for water quality problems. At both sites, water samples are analyzed for chlorine levels, pH, organic and inorganic pollutants, bacteria, and odor, among other things. DEP also monitors a few groundwater sources which provide supplementary water to certain areas of the city, including Laurelton, Queens Village, and Cambria Heights.

Until now, New York City—unlike most large cities across the country—has not been required to filter its water. (Filtration involves passing the water through filters made of layers of sand, gravel, and charcoal that can remove very small particles.) However, in a historic agreement reached between the city and the federal Environmental Protection Agency in 1997, the city agreed to build a filtration plant for the Croton watershed system—the oldest of the three, and that which has been most affected by suburban growth. In return for a commitment to expand the city's watershed protection program by funding improvements to the affected upstate communities on the west side of the Hudson, the city was granted a waiver from the requirement to filter water from both the Catskill and Delaware systems.

Manhattan Water Sampling Locations

Water Sampling Stations
In the late 1990s, over 800 sampling stations were installed throughout the city in what were considered representative locations. Made of heavy cast iron, the stations stand roughly four feet tall. Inside a protective casing, a small copper tube conducts water from a nearby water main into a spigot that runs into a small sink. Each month, the department collects between 800 and 900 samples from roughly 500 locations.

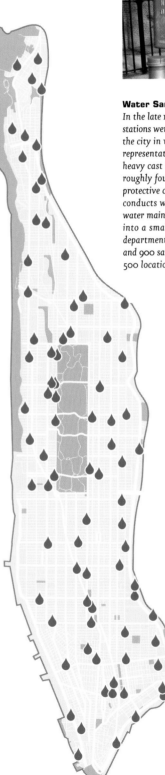

industrial users—includes human waste, food scraps, oil, soaps, and chemicals. And because New York City is one of the few cities in the country that does not have separate systems for handling storm water and wastewater, it also includes runoff from rain and storms.

New York City's sewer system is among the most extensive in the nation. Its 6,600 miles of sewer

The city's earliest sewer dates back to Dutch colonial times, when a channel was dug down the middle of Broad St. in lower Manhattan and decked over. But through

mains and pipes—despite their advanced age—are both functional and generally reliable. Together with 14 wastewater treatment plants, they handle and process 1.3 billion gallons of sewage per day.

The wastewater handled by the system—from showers, sinks, bathtubs, toilets, washing machines, dishwashers as well as

the middle of the nineteenth century, there really was no sewer system to speak of in the city: residents and businesses disposed of waste in backyard outhouses or simply dumped it into the gutters along the streets. In 1849, after a series of deadly cholera outbreaks, the city initiated a sewer-building program. Over the next 50 years, the network would extend to all developed sections of the city, and even tenement houses would begin offering flush toilets.

Sewage

But the water moving through the system and out into the waters surrounding the city remained largely untreated. The first modern sewage treatment plant in the country was constructed in the late nineteenth century for the then city of Brooklyn: several small facilities built in Coney Island permitted solid matter to settle to the bottom of tanks before being removed and buried, while simultaneously treating the liquid with chlorine before returning it to the ocean. But the need for sewage treatment plants was not widely accepted until 1931, when the city finally

Construction of the city's sewers, which were largely made of brick, began in 1865 under the authority of the Croton Aqueduct Department.

published a comprehensive plan to construct modern sewage treatment plants. It would be another 50 years before the plants envisioned in that plan were completed and online.

With the advent of treatment plants, the solid matter—or sludge—could be separated from the water, which was treated with chemicals and returned to the ocean. But throughout most of the twentieth century, the separated sludge was also dumped in the ocean—initially at a site 12 miles from the Jersey shore and subsequently farther out, at a dump site 106 miles offshore. At the time the city finally stopped ocean dumping, in 1991, it was disposing of over six million tons of sludge at its ocean dump site annually.

The "Poo-Poo Choo-Choo" For a decade after Congress prohibited ocean dumping of sludge in New York Harbor in 1992, New York's sludge was transported by train over 2,000 miles to a small outpost in west Texas called Sierra Blanca. There, 90 miles southeast of El Paso, roughly 250 tons a day of waste would be spread atop the fields, at one of the biggest sludge dumps in the world. The locals welcomed the sludge, and found ways to live with the smell when the breeze came from the north: one said, "It smells like a hog farm," and another, "To me, it smells like money." The last train called at Sierra Blanca in 2001, by which time the Department of Environmental Protection had found cheaper alternatives for sludge disposal.

Ocean Dumping Sites *Until the early 1990s, the city dumped its sewage—initially raw and subsequently processed—into the ocean at two sites. The initial site was only 12 miles off the coast of New Jersey, in water depths of 88 feet. A subsequent site was located 106 miles out into the open ocean, off the Continental Shelf, and featured water depths of about 7,500 feet.*

Sewage

Collection System   New York's sewer system includes over 6,000 miles of sewer pipes, ranging in size from six inches to more than 89 inches in diameter, as well as 145,000 catch basins (stormwater drains) and 5,000 seepage basins (catch basins that allow water collected to pass into the ground). City sewer pipes are generally buried more than 10 feet underground—below the level of clean water pipes, so as to avoid any possible contamination if they leak.

The pipes themselves were installed at varying points in history and are constructed of a range of materials. Some of the oldest, located in parts of Brooklyn and Manhattan and dating as far back as 1851, are made of cement, brick, and clay. Reinforced concrete and iron can be found on some of the more modern and larger pipes. Roughly two-thirds of the system, however, consists of vitreous clay pipes—which are largely impervious to the chemical actions of the sewage.

In general, sewage from several buildings is carried through separate pipes into a large pipe called a lateral. The laterals are connected first to a submain, then to a main pipe, and eventually to what is known as an "interceptor," which feeds directly into the treatment plant. Gravity plays a large role in keeping the sewage moving, with a target rate of two to three feet per second.

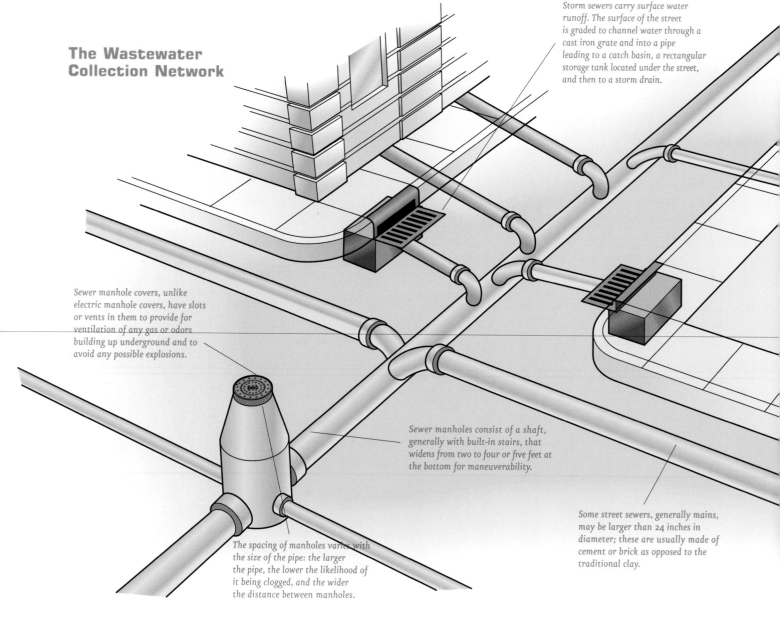

The Wastewater Collection Network

Storm sewers carry surface water runoff. The surface of the street is graded to channel water through a cast iron grate and into a pipe leading to a catch basin, a rectangular storage tank located under the street, and then to a storm drain.

Sewer manhole covers, unlike electric manhole covers, have slots or vents in them to provide for ventilation of any gas or odors building up underground and to avoid any possible explosions.

Sewer manholes consist of a shaft, generally with built-in stairs, that widens from two to four or five feet at the bottom for maneuverability.

The spacing of manholes varies with the size of the pipe: the larger the pipe, the lower the likelihood of it being clogged, and the wider the distance between manholes.

Some street sewers, generally mains, may be larger than 24 inches in diameter; these are usually made of cement or brick as opposed to the traditional clay.

The diameter of building pipes depends upon the number of plumbing fixtures in the building as well as the grade or inclination of the pipe.

Sewage from buildings leaves through pipes that connect with the street sewer, which is a combination sanitary and storm sewer—generally 12 inches or more in diameter—that is typically made of clay.

In large apartment buildings, sewage lifts involving air compressors are used to raise the level of the waste from a building cellar to the street conduits.

Septic Tanks

Some homes in New York City, particularly in more remote areas of the outer boroughs, still rely on septic tanks, which function like tiny private sewage treatment plants. Made of concrete, steel, or fiberglass, the tanks are generally buried in a yard and hold large volumes of wastewater. Waste flows into the tank from sinks, toilets, and showers and is attacked by bacteria that break down the organic material. New water flowing into the tank pushes water already there out into a drain field, which consists of perforated pipes buried in trenches in the yard, and from here water is absorbed and filtered into the ground.

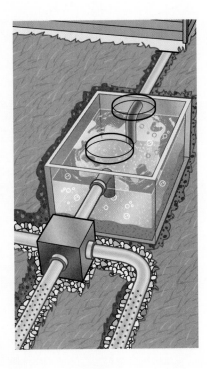

Sewage

Combined Sewer Overflow New York is one of only 800 or so cities in the country that rely on what's known as a "combined sewer system"—one that mixes storm water with wastewater, and sends them both on to the same treatment plant. During dry weather, the combined system poses no problems: the treatment plants handle waste from the sanitary sewers. However, if rain results in run-off significant enough to more than double the "dry-weather flow" into the plant, the system risks backing up a mix of storm water and raw sewage into homes and streets. To avoid this, all excess flow—above the level that the plant can handle—is diverted to a combined sewer overflow (CSO) outfall and discharged, untreated, into the harbor.

There are over 700 CSO outfalls in the harbor, about 450 of them in New York City alone. And they are used frequently: overflow occurs about half the time it rains, leading to an estimated 40 billion gallons of untreated waste (20 percent of which is raw sewage) pouring into the city's waterways. To combat the problem, the city is currently building three underground reservoirs—two in Queens and one in Brooklyn—that can hold excess water until water levels subside and the overflow can be pumped to treatment plants.

Anatomy of a CSO

The CSO system depends heavily on the operation of two particular pieces of machinery: regulators and tide gates.

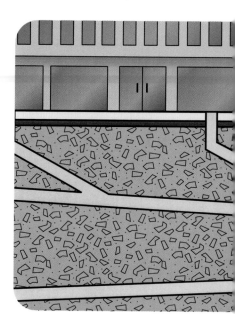

CSO Locations **The city's shoreline is dotted with CSOs, or combined sewer overflows, which release into the harbor excess runoff—including, at times, untreated sewage—at times of heavy rainfall. Each of these outflows is registered with the New York State Department of Environmental Conservation.**

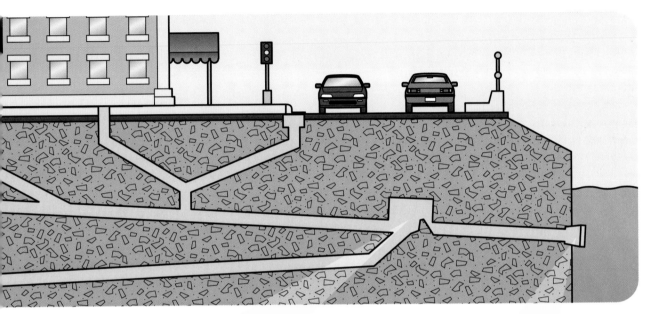

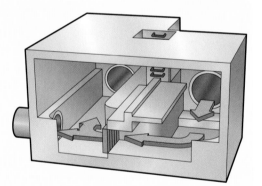

Regulators *are devices used to control the flow of wastewater to the treatment plant during dry weather and to both the treatment plant and the outfall pipes during wet weather. In general, roughly twice the normal flow will be diverted to the plant and the remainder discharged through tide gates and into receiving waterways through outfall pipes.*

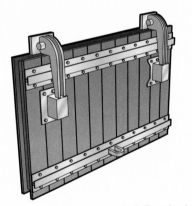

Tide gates *are simple flaps placed at the end of the outfall pipe, which remain closed until water pressure inside the pipe exceeds the water pressure (and the weight of the flap) in the outside waterway. The gates, made of cast iron, timber, or stainless steel, prevent saltwater from entering the sewer system. More than 500 of these gates are currently in operation around the city.*

Sewage

Catch Basins

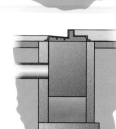

Catch basins collect runoff while keeping larger items—such as trash and small children—out of the sewer system. Rectangular catch basins are common in New York, although other shapes may be found, and walls are lined with either concrete or brick. It is common to find up to six catch basins at a particular intersection.

Floatables Residents of the metropolitan region are all too familiar with what is generally referred to as "floatables"—plastic, paper, and Styrofoam items that accumulate along the shoreline or on beaches in the region after a storm event. Much of what accumulates in this fashion tends to be street litter, which has been swept into storm drains and sewers and traveled out through a CSO outfall during a storm.

Several initiatives are under way to reduce the volume of floatables in the harbor. One is the installation of hoods at the city's 130,000 catch basins located on city streets. The hood serves as a baffle to prevent floatables that fall into the basin from ever moving into the sewer; it also helps prevent sewer gases from reaching the street. Although it requires the regular cleaning of built-up debris in the catch basin by DEP or its contractors, the installation of hoods has been shown to reduce the amount of litter entering the sewer from an individual catch basin by 70 to 90 percent.

Catch-basin hooding *The city's catch-basin hooding program involved more than just placing a hood at each location. Each catch basin was inspected, inventoried, and electronically mapped before the hood was installed. The database and maps that resulted from those inspections now serve as a management tool to direct catch-basin repair and maintenance activities.*

In addition, DEP has installed floating barriers or booms at 23 locations to capture floatables from major CSO outfalls. Skimmer vessels are used to remove floatables from within the boomed areas. Four belt-type skimming vessels, each 45 feet in length, are used to conduct the skimming operation: the *Ibis*, the *Piping Plover*, the *Green Heron*, and the *Snowy Egret*. The boats have "wings" that can open and close to catch litter, a front conveyor to bring debris into a holding area, and a rear conveyor to deposit it into a barge. They tend to pick up about five cubic yards of debris a week, filling a barge with refuse roughly twice a month.

In addition, the city relies on a special skimmer vessel, the *Cormorant*, for its open-water skimming program. The *Cormorant* uses a net rather than a belt to collect floatables and is therefore able to collect timber and other heavy waterborne debris. Twelve years old and 120 feet in length, she is able to hold 24 tons of debris.

Booming and Skimming Operations

Booms or floating barriers have been installed at about two dozen sites in the harbor to contain floatables passing out of combined sewers. Skimming operations attempt to collect any floatables not contained by the booms.

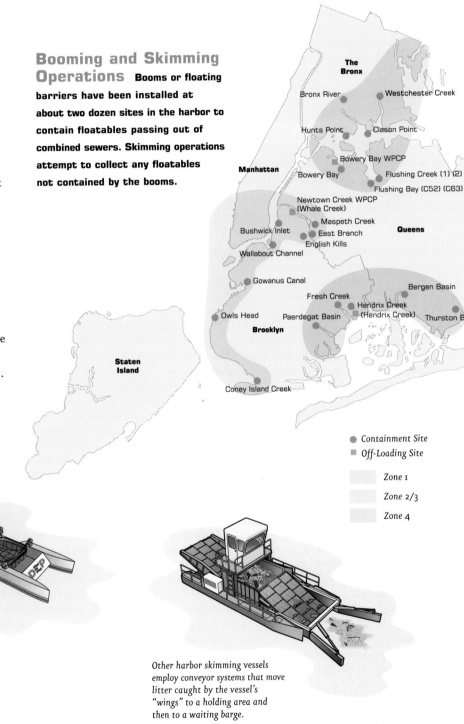

● Containment Site
■ Off-Loading Site

Zone 1
Zone 2/3
Zone 4

The New York City Cormorant, the largest in DEP's fleet of harbor-skimming vessels, was commissioned in 1993 and built at the Amfels Shipyard in Brownsville, Texas. It features a net to collect larger debris in open water.

Other harbor skimming vessels employ conveyor systems that move litter caught by the vessel's "wings" to a holding area and then to a waiting barge.

Sewage

Sewage Treatment Day in and day out, New York City relies on 14 sewage treatment plants and approximately 100 pumping stations to keep its wastewater moving. The location of the sewage treatment plants is based largely on the city's topography: they are located at the lowest possible elevations, to allow gravity to move sewage to the plants. Pumping stations, which push sewage along where gravity is not sufficient to keep it moving, are required primarily to serve low-lying areas.

Each treatment plant is fed by a master sewer, or "intercepting sewer," which collects from the various sewer mains within the district. Not all plants are alike, however: only eight have dewatering facilities, which allow comprehensive treatment of the sludge; the others produce a sort of wet, soillike substance which must be moved, by sludge boat, to a sister plant for dewatering. The newest plant, the North River facility, is typical: sludge from the plant is transferred by boat for dewatering at the Wards Island wastewater treatment plant. Once dewatered, the products remaining— generally known as "biosolids"— are removed by companies that have long-term contracts with the city for beneficial reuse of the sludge.

How a Treatment Plant Works

1. Upon entering the plant, wastewater is pumped through bars or screens to remove large debris such as newspapers, sticks, rags, and cans.

2. Sewage pumps lift the wastewater to primary settling tanks at surface level.

3. Heavier materials settle to the bottom of the tank, while lighter materials—as well as oil and grease —float. Collection devices skim oil and floatables from the top and remove the primary sludge that has settled at the bottom.

4. The partially treated wastewater flows to a secondary treatment system (known as the "activated sludge process"), where air and "seed sludge" from the plant are added to the wastewater to break it down further. Air pumped into the aeration tanks stimulates the growth of bacteria and other organisms that consume much of the remaining organic materials that pollute the water.

5. The aerated water moves to final settling tanks, where additional settlement occurs. These solids are removed and join the primary sludge for further processing. The water, after being treated with chlorine, is released as effluent into a nearby waterway.

6. Thickeners are added to the sludge, while additional water is removed from it and treated. The material sits for a day to allow the solids to become more concentrated.

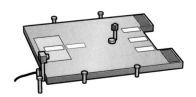

7. The solids are sent to a final set of oxygen-free tanks, called digesters, which are heated to 95 degrees Fahrenheit to stimulate the growth of anaerobic bacteria, which will consume the organic material in the sludge over a period of 15 to 20 days.

Sewage Treatment Areas and Plants

(capacity in millions of gallons per day)

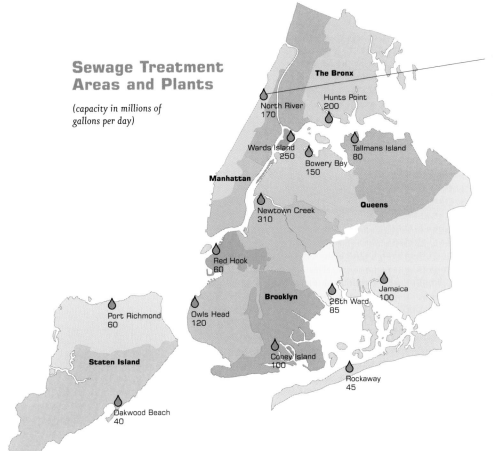

The Bronx

North River
170

Hunts Point
200

Wards Island
250

Tallmans Island
80

Bowery Bay
150

Manhattan

Newtown Creek
310

Queens

Red Hook
60

Brooklyn

26th Ward
85

Jamaica
100

Port Richmond
60

Owls Head
120

Coney Island
100

Staten Island

Rockaway
45

Oakwood Beach
40

The North River Treatment Plant
Until 1986, untreated sewage from the West Side of Manhattan flowed directly into the Hudson River. Sites for a treatment plant had been identified as far back as 1914, but it was not until 1962 that the current site—between 137th and 145th Sts., west of the West Side Highway—was selected and approved by the City Planning Commission. Design and construction took a full 24 years, in part because of the plant's unique physical configuration: built on a 28-acre reinforced concrete platform over the Hudson River, it sits on 2,300 caissons drilled into bedrock up to 230 feet beneath the river. On its roof is Riverbank State Park, a 28-acre recreational facility that includes pools, a skating rink, an athletics center, and sports fields. The facility handles 125 million gallons of wastewater each day in dry weather, and up to 340 million gallons per day when it rains.

Sludge Barges Municipal sludge vessels have been a common sight in New York Harbor since the late 1930s, when they began carrying sludge to disposal sites outside the harbor. The boats were named after the treatment plant they served (e.g., *Wards Island*, *Tallman Island*, *Coney Island* were the first) and had a total capacity of 40,000 cubic feet apiece. The fleet changed only slightly over the next 50 years, as older boats were retired and several new ones brought on line. In 1987, however, when the ocean disposal site was moved from 12 to 106 miles offshore, a new fleet of four oceangoing barges was purchased and the motorized vessels were relieved of dumping duties.

Today, three of the old motorized vessels are used to transport over 300,000 cubic feet of sludge each day within the harbor, from plants without dewatering facilities (Owls Head, Rockaway, Newtown Creek, and North River) to those with them (Wards Island, Hunts Point, and 26th Ward). Each sludge vessel has four crews of six persons, who work 12-hour shifts. Two vessels are used on a six-day schedule, with the third on standby—making roughly 1,200 trips each year.

Sewage

Sludge Processing   Since 1991, when it ceased dumping sludge into the ocean, New York City's Department of Environmental Protection has had a variety of contracts with private companies to dispose of the city's sludge, or biosolids:

- Roughly half of all biosolids are processed into fertilizer pellets at a plant in the Hunts Point section of the Bronx. Although the pellets are sold to users across the country, the lion's share are sent south, to fertilize citrus groves in Florida.
- A second contract provides for direct application of biosolids to grazing land and cornfields in Virginia and grazing land and wheat fields in Colorado.

- A third contract combines alkaline stabilization (lime treatment) of biosolids at a plant in New Jersey, pelletization at a plant in Arkansas and composting in Pennsylvania. The products of these processes are respectively sold in New York and New Jersey as fertilizer for hay and corn crops, in Arkansas as a blended fertilizer, and in Pennsylvania as a composted topsoil blending material.

The city itself uses biosolids produced at its waste treatment plants as an additive to soil within the five boroughs. Biosolids compost has been used along the Major Deegan in the Bronx, at the Queens Botanical Garden, at a residential building in Manhattan, and even on the lawns of Gracie Mansion.

From Sludge to Fertilizer Roughly half of the city's biosolids presently move to a pelletization facility in Hunts Point operated by the New York Organic Fertilizer Company, a subsidiary of Wheelabrator Technologies.

1. Biosolid materials move to the plant via dump trucks or barges. Once at the plant, the material is taken to a "tipping area," where it is dumped into pits for storage.

2. First, it goes to the pin mixer, where it is mixed with recycled pellets from previous batches.

3. It then moves to a drier drum, where an air heater heats it to between 600 and 1,000 degrees Fahrenheit, evaporating any water. The material goes through the drum three times, eliminating any pathogens.

4. Hot air and pellets come out of the drum at about 180 degrees Fahrenheit and are separated.

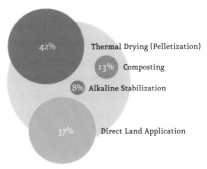

42% Thermal Drying (Pelletization)

13% Composting

8% Alkaline Stabilization

37% Direct Land Application

The Afterlife of Sludge *New York City sludge, often referred to as "biosolids," is no longer dumped in the ocean. Instead, it is used primarily to fertilize crops and improve soil conditions for plant growth: its application increases the soil's ability to hold water, stimulates root growth, and generally improves the texture of the soil.*

8. The air stream is then sent to an oxidizer, which helps destroy carbon monoxide, and is heated to 1,650 degrees Fahrenheit, which destroys remaining odors. Once cooled, the air can be sent out a chimney stack.

7. The air stream moves to a cyclone, where dust is removed, and then to a wet scrubber, where ammonia is removed.

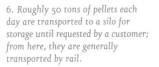

5. The pellets move to a vibrating screen where a sieve separates different-sized pellets and removes over- and undersized ones.

6. Roughly 50 tons of pellets each day are transported to a silo for storage until requested by a customer; from here, they are generally transported by rail.

Sewage

Managing Effluent Most New Yorkers are aware that the waters that surround the five boroughs have gotten cleaner over the past few decades. Fish species largely absent from the harbor for years, such as striped bass and bluefish, are now regularly spotted and harvested by recreational fishermen. Competitive swims now take place in the waters of the Hudson. And to many, the water just looks cleaner: more rarely are plastic bottles or aluminum cans to be found bobbing along the shoreline.

The improvement in water quality is a direct outgrowth of the tremendous investment in wastewater treatment plants that has been made over the last 50 years in New York. And this investment continues, in an attempt to reduce the number of incidents that introduce organic matter into the region's waters (and hence minimize the volume of bacteria which, in feeding on this matter, reduce all-important oxygen levels in the water). But even today, major sewage discharges or rain-related runoff can pose tremendous risks to aquatic life—particularly to organisms living near the bottom of the harbor, and particularly in conditions of sustained summer heat.

Certain spots within the harbor have proved particularly problematic with respect to oxygen depletion, primarily western Long Island Sound and parts of Jamaica Bay. Among the most difficult problems to solve has been that of the Gowanus Canal, a former commercial waterway in Brooklyn that for years ranked among the most polluted waterways in the nation. Reactivation of its Flushing Tunnel, first installed in 1911 to draw dirty water out of the canal and allow aerated water from the nearby Buttermilk Channel to flow into it, was undertaken in the late 1990s and has helped raise the level of dissolved oxygen in the canal's water. Work remains to be done to improve quality further, and city DEP and the Army Corps of Engineers are collaborating on a broader set of remedies.

In addition to problems with specific hot spots, unexpected events leading to sewage discharge can also create depleted oxygen conditions in the waters surrounding the city. Mechanical breakdowns at treatment plants are relatively rare but do occur; blackouts are equally problematic, due to the treatment plants' reliance on electric pumps. During the Northeast blackout in August 2003, untreated waste flowed into the East River from a pumping station on East 13th St. Without working pumps, water pressure pushed open the tide gates that control sewage discharge—and an estimated 145 million gallons of raw sewage escaped. During the same incident, another 345 million gallons poured into the Hudson River and into waters off Brooklyn from other treatment plants.

The Nitrogen Control Program

Nitrogen discharge from the city's waste treatment plants, and its impact upon water quality in Long Island Sound, has proved an enormous and costly problem over the last decade. Two lawsuits for improper operation of DEP plants were filed in 1998—one by the New York State Department of Environmental Conservation and another by the Long Island Soundkeeper and the state of Connecticut. Since that time, the city has taken a series of measures—including retrofitting certain plants—to reduce the level of nitrogen discharge that flows into the Upper East River and Jamaica Bay.

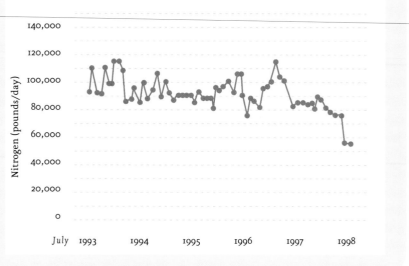

Upper East River Wastewater Pollution Control Plants
Monthly total nitrogen discharge

Nitrogen (pounds/day)

140,000

120,000

100,000

80,000

60,000

40,000

20,000

0

July   1993   1994   1995   1996   1997   1998

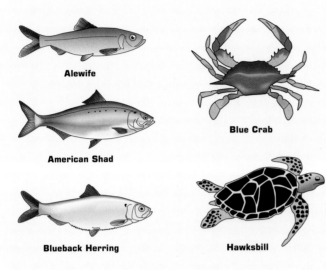

Alewife

American Shad

Blueback Herring

Blue Crab

Hawksbill

Return of the Natives

As water conditions in New York Harbor continue to improve, and animal protection programs mature, multiple species have begun to return to the harbor. These include sea turtles, striped bass, bluefish, and American shad, among others. Harbor seals have recently been spotted sunning themselves on Manhattan's West Side docks.

Monitoring Harbor Water Quality

As far back as 1909, with the establishment of the New York Water Quality Survey by the Metropolitan Sewerage Commission, New York has carefully documented the cleanliness of its waters. Today, water quality stations at 53 locations in the harbor monitor surface and bottom water quality and analyze the levels of pollutants in samples of sediment taken from the bottom. In addition, the Department of Environmental Protection's HSV *Osprey*, a 55-foot custom-made boat complete with a shipboard laboratory, undertakes year-round water quality sampling to meet federal requirements.

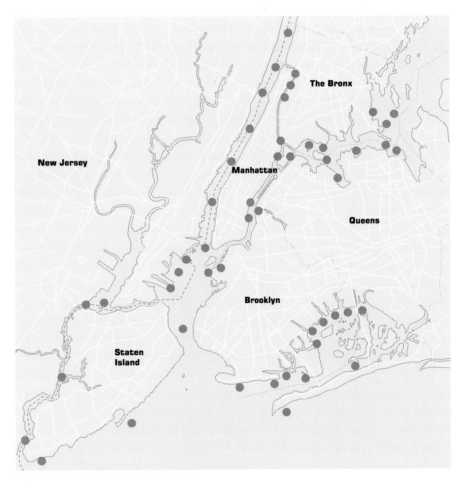

● Water quality station

Keeping New York clean is a Herculean task, which falls primarily to the city's Sanitation

Department and its 10,000 employees. Responsible for everything from garbage collection to street cleaning to snow removal, the department represents the world's largest sanitation workforce— and one of the busiest. Each day, the city produces about 12,000 tons of residential (from homes) and municipal (from schools and other city institutions) waste; twice a week, and in some areas three and four times a week, it is the department's job to collect it. It also collects recyclables from these same areas once a week, empties litter baskets on street corners, and runs a variety of other specialized and seasonal waste collection programs for city residents (collecting Christmas trees, yard waste, electronic and hazardous waste, for example).

One thing the department does not do is collect commercial waste from businesses. Since 1957, the collection of commercial waste —from offices, restaurants, stores, and factories—has been left to private haulers and commercial waste companies. The city's role is limited to licensing haulers and granting permits to, as well as inspecting, the transfer stations that they rely on for out-of-state export.

This division of responsibility between private and public sectors is relatively new. Historically, waste disposal fell either to private or public sectors—with the latter supplanting the former in the late nineteenth century. By that time, the city's streets were overwhelmed with a fetid combination of human and animal waste, ashes, and accumulated street dirt.

Garbage

New York City's Garbage Timeline

1865 — 1880 — 1895 — 1910 — 1925

1866: New York City's Metropolitan Board of Health declares war on garbage, banning the "throwing of dead animals, garbage or ashes" into the streets.

1872: New York City stops dumping its garbage from a platform built over the East River, and instead takes it out on barges and dumps it into the ocean.

1881: The Department of Sanitation, originally known as the Department of Street Cleaning, is founded, taking over responsibilities for waste management from the Metropolitan Board of Police.

1885: The nation's first garbage incinerator is built on Governors Island. Hundreds of other incinerators will spring up elsewhere in the country in the decades that follow.

1896: New York City requires residents to separate household waste—organic waste in one receptacle, ash in another, and dry garbage in a separate bag.

Horses alone contributed 500,000 pounds of manure and 45,000 gallons of urine each day. Although private property owners hired laborers to clean their own areas and dump their waste in the ocean, the tenement districts on both East and West sides of Manhattan remained untouched. What couldn't be eaten by the family pig, if the family was fortunate enough to have one, was simply dumped on the street.

The Blizzard of 1888—the very same storm that led to the placement of utility lines underground—provided momentum for the idea of a regular, accountable sanitary force within the city. Seven years later, the existing Department of Street Cleaning fell into the hands of the legendary Colonel George Waring, who instituted organizational reforms that would lay the foundation for the comprehensive waste management system that exists in the city today.

A SECTION FOREMAN WITH HIS SWEEPERS READY TO MARCH.

Waring and the White Wings

No figure is more prominent in New York City's garbage history than Colonel George Waring, a military man who—in his short tenure at the helm of the Department of Street Cleaning, from 1895 to 1897—transformed not only the department but the profession as a whole. Waring introduced new techniques and technology, reshaping the organization and instilling a sense of pride and professionalism across the department's workforce. He issued crisp white uniforms to all sanitation workers, who quickly became known as "the White Wings," and amassed a following among the public as the personification of cleanliness and hygiene. (That tradition continues today: the department encourages shined shoes, closely cropped facial hair, nails no more than 1/4 inch from the end of the finger and a minimum of jewelry.) In addition to introducing a highly ambitious recycling program, which would last more than a decade, he developed both a school for street cleaning and an organization of juvenile street cleaning leagues, made up of children whose job it was to promote good hygiene among poorer residents of the city.

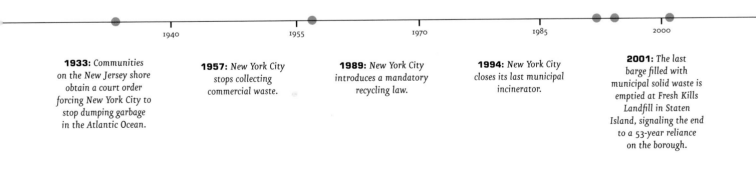

1933: Communities on the New Jersey shore obtain a court order forcing New York City to stop dumping garbage in the Atlantic Ocean.

1957: New York City stops collecting commercial waste.

1989: New York City introduces a mandatory recycling law.

1994: New York City closes its last municipal incinerator.

2001: The last barge filled with municipal solid waste is emptied at Fresh Kills Landfill in Staten Island, signaling the end to a 53-year reliance on the borough.

Garbage

Sanitation Equipment

The Department of Sanitation operates out of 59 districts, divided primarily along community district lines, and relies on a fleet of close to 6,000 vehicles. Roughly 2,100 of these are frontline collection trucks, most of them rear-loading vehicles with a 25-cubic-yard capacity, 10 tires, and a crew of two. In addition, the department has 450 mechanical street sweepers, 350 salt/sand spreaders, 275 specialized collection trucks, 280 front loaders, and over 2,000 other support vehicles. Most of these vehicles are replaced on average every seven years; for that, the department allocates between $100 million and $200 million each year.

Repair and maintenance occupies a huge amount of the department's time and resources. All vehicles are serviced every 30 days (the department goes through an average of 100,000 tires each year!). The department's Central Repair Shop, in Maspeth, Queens, is said to be the largest nonmilitary shop in the country—1.26 million square feet spread over six floors. In addition to a body shop, special chassis shop, minor and major component rebuilding shops, tire shop, machine shop, radiator shop, upholstery shop, and forge/tractor shop, it includes a dynamometer room, where engines and pumps are run and tested before they are put into the trucks. The department also maintains decentralized district and borough shops for more routine maintenance.

Sanitation's trucks are among the least polluting in the nation. The entire fleet has been converted to ultra-low sulfur diesel fuel, well in advance of the federal mandate for commercial vehicle conversion. Simultaneously, the department is testing diesel particulate filters on over 100 trucks; these filters, when used in combination with the ultra-low sulfur diesel fuel, can reduce emissions by roughly 90 percent.

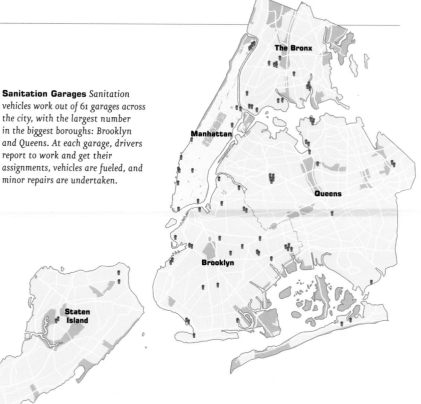

Sanitation Garages *Sanitation vehicles work out of 61 garages across the city, with the largest number in the biggest boroughs: Brooklyn and Queens. At each garage, drivers report to work and get their assignments, vehicles are fueled, and minor repairs are undertaken.*

Types of Litter Baskets

Over 25,000 litter baskets can be found at designated locations throughout the city. Roughly 2,000 of them are purchased by Business Improvement Districts (BIDs); the remainder are provided by the Department of Sanitation. The department collects from these baskets anywhere from two to seven times each week, with heavy commercial areas allocated daily pickups. Three types of litter baskets are currently in use:

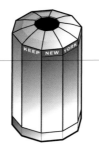

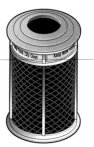

Corcraft basket *The Corcraft basket was selected as the department's standard litter basket in 2000. It has eight welded vertical support braces, which are laid over perforated steel sheeting with a hole pattern design. The basket has a reflective decal around the rim to provide added visibility at night. Weighing in at only 33 pounds, it is also cheap, costing under $100.*

Forms and Surfaces basket *The Forms and Surfaces basket, also known as "the California basket," has been in use since 1997 in more than a dozen heavily trafficked commercial and tourist areas. Weighing close to 270 pounds, it has a convex lid, perforated pattern side panels, and a side-door opening, and costs approximately $700.*

Victor Stanley basket *The Victor Stanley basket, introduced in 1997, is made entirely of steel. Weighing 230 pounds and holding 45 gallons, it features a side-door opening and a convex lid and costs $700.*

The Sanitation Fleet

Meet the Rear
Loader Most collection trucks
are rear-loading packer trucks.
Trash goes into the "hopper." As
more garbage is added, it
pushes against a bladelike panel
that is backed by 2,200–2,500
pounds of pressure. As the
garbage pressure increases, the
panel backs up slowly until the
truck is full. The truck is unloaded
by backing up to the dump
location, opening the back
(including lifting the blade), and
then tilting the body of the
truck so that garbage can be
pushed out with an ejection blade.
On an average day, a standard
rear-loader truck with a 25-cubic-
yard capacity will collect 12–14
tons of garbage. The average life
span of a collection truck is seven
years, assuming it is serviced
every 30 days.

Dual bin collection truck
*Designed to collect metal, glass,
and plastic in one bin and paper in
another, these Mack trucks hold
25 cubic yards of material. Slightly
more capacity (60 percent)
is allocated to paper than to metal,
glass, and plastic (40 percent).*

Salt/sand spreader *Most salt
spreaders carry 16 cubic yards of salt
or sand, which is sprayed onto
the roadway in varying amounts by a
spinning disk in the back of the truck.*

Roll on/Roll off chassis *Also called
tilt-body container trucks, these
vehicles are used for containerized
collection. Equipped to carry
containers that range from 20 to 40
cubic yards, their chassis are
equipped to tilt the container and slide
it off the bed of the truck as the
truck moves forward.*

Tilt–body truck *These trucks are
used to transport broken-down
vehicles to a repair facility. They are
also used to remove abandoned or
illegally parked vehicles.*

Snow plow *During snow falls,
regular collection trucks are "dressed"
with plows and sent out to remove
snow from the streets. Collection
trucks are purchased with a
mechanism to lift and tilt the plow.*

Hoist compactor (EZ pack)
*Also known as front-loading packer
collection trucks, these vehicles
use a front-mounted lifting arm to
service containers with capacity
from two to eight cubic yards.*

Mechanical broom *More commonly
called street sweepers, these vehicles
have two gutter brooms and
one main underbody broom that sweep
litter and debris from the gutter
line of the city's streets. Although they
usually spray water as they
sweep, these vehicles can work without
water during drought emergencies.*

Cargo van *Also called a
mechanic's rig, this vehicle is used
to carry equipment for
vehicle repair at remote sites.*

Garbage

Sample Collection Route: Daily and Weekly Within one district in Brooklyn, areas will receive pickup at different times of the day and on different days of the week.

Weekly Refuse Collection

- *Biweekly*
- *Triweekly*
- *Commercial strips*
- *Industrial strips*

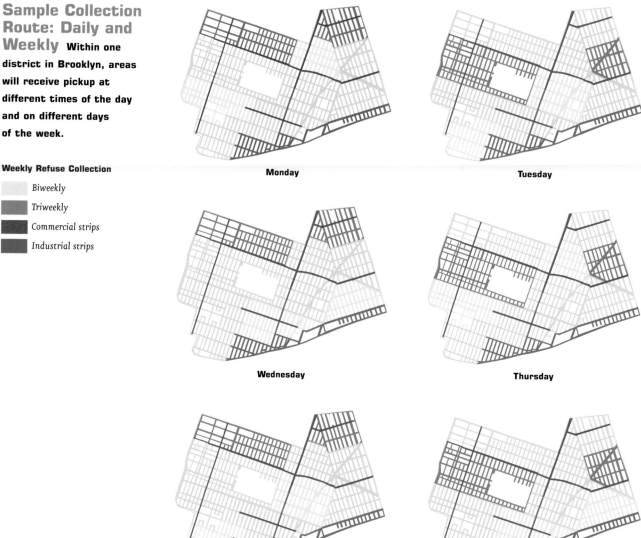

Monday

Tuesday

Wednesday

Thursday

Friday

Saturday

Collection Routes Garbage collection routes in New York City are determined by a variety of factors: labor union agreements, truck capacity, and operational issues—such as congestion, curb accessibility, and topography weigh most heavily. While most residential areas are served by two collections per week, more densely populated areas receive three pickups. The New York City Housing Authority is serviced four times per week, and New York City public schools—1,100 of them—receive daily service.

In addition to traditional garbage collection, the department collects recyclables on a weekly basis across the city. Specialized collections are done for autumn leaves, during roughly six weeks in the fall in areas where gardens are commonplace, and for Christmas trees, which are collected from curbside and taken to either a Department of Parks facility on Randalls Island and chipped into a mulch or to Fresh Kills, where they are composted.

Special Events New Yorkers love special events, and not a year passes without many of them: New Year's Eve in Times Square, the Thanksgiving and St. Patrick's Day parades, and occasionally a ticker-tape parade celebrating a local sports team's victory. For each event, Sanitation—with the help of the Police Department—estimates how many people will attend. When it is over, workers with leaf blowers on their backs move in to blow trash to areas where it can be collected. Front-end loaders come in to scoop up the piles of trash, supplemented by workers with hand brooms. Street sweepers come in once the initial collection is complete.

After the Ball Drops The annual Times Square New Year's Eve celebration offers among the most challenging cleanups of the year for the Sanitation Department. Sanitation staff—78 of them—begin gathering at 11 p.m., and head out for action soon after midnight. The cleanup, which involves collection trucks, mechanical brooms, hand brooms, wreckers, and a box truck, lasts all night and through the next day.

Garbage

Exporting Garbage With no landfills or incinerators operating within city limits, nearly 12,000 tons of residential waste must be exported from New York City each day. Much of it leaves by truck—not by the Sanitation vehicles which are a familiar sight on city streets, but by 18-wheeled transfer trailers that carry it to landfills in Pennsylvania, New Jersey, and Ohio. Some 66 transfer stations, nearly all of them in the outer boroughs, serve as the lynchpin of this trucking network: sanitation vehicles bring trash to these privately operated stations, where it is loaded onto larger trucks for long-distance travel to remote landfills.

There are exceptions, primarily in Manhattan. Roughly two-thirds of Manhattan's residential waste moves by Sanitation Department vehicles to the Essex Resource Recovery facility in Newark, New Jersey, where it is burned to generate electricity. Additional Manhattan waste, and some from Staten Island and Queens, currently moves via New Jersey-based transfer stations to landfills. And much of the Bronx's residential waste is containerized at Harlem River Yards, and moves from there by rail to a landfill in Virginia.

The current truck-based system, highly unpopular among environmentalists and the communities near transfer stations, resulted from the relatively swift closure of the city's barge-fed landfill at Fresh Kills on Staten Island and the abandonment of the network of marine transfer stations that supported it. Though the landfill arguably had capacity to operate a decade or so longer, its closure in 2001—prompted by political considerations—left the city wholly reliant on the private sector for disposal and with no alternative to long-distance trucking for much of the residential waste stream.

Where Does the Garbage Go?

Each borough relies on a number of private vendors to dispose of the city's waste, primarily outside of the state. Although some is sent to the Essex Resource Recovery Facility in Newark, the bulk of the city's garbage ends up in landfills in Pennsylvania and Virginia.

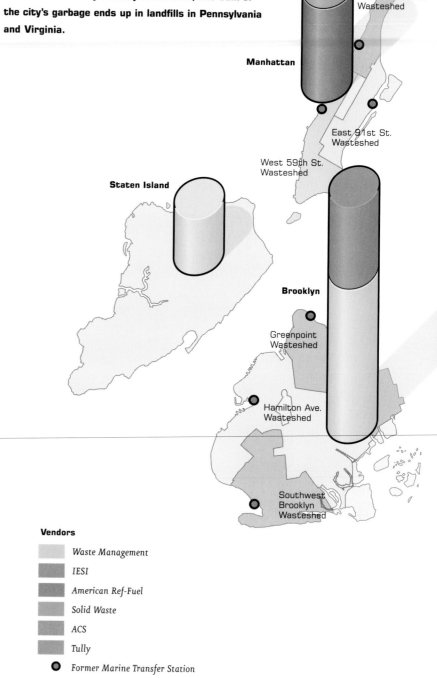

West 135th St. Wasteshed

Manhattan

East 91st St. Wasteshed

West 59th St. Wasteshed

Staten Island

Brooklyn

Greenpoint Wasteshed

Hamilton Ave. Wasteshed

Southwest Brooklyn Wasteshed

Vendors

- Waste Management
- IESI
- American Ref-Fuel
- Solid Waste
- ACS
- Tully
- ● Former Marine Transfer Station

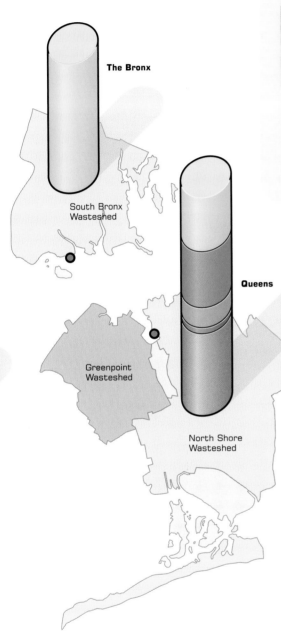

The Bronx

South Bronx
Wasteshed

Greenpoint
Wasteshed

Queens

North Shore
Wasteshed

The Garbage Train Each day, a train made up of 35 cars of containerized garbage pulls out from a warehouse at the Harlem River Yards in the South Bronx, destined for a landfill in Waverly, Virginia, some 400 miles away. Inside the green waste containers is much of the residential waste produced in the Bronx as well as a significant amount of commercial waste produced in the city. Though hauled by CSX Railroad, both the containers and the landfill they are traveling to belong to Waste Management, the largest waste company in the country and one of the city's two export service providers in the Bronx.

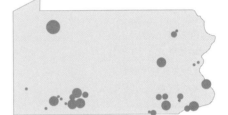

Pennsylvania *Almost two-thirds of the city's municipal solid waste ends up in one or another of two dozen landfills in Pennsylvania. Because landfill capacity there is dwindling, Pennsylvania's prices are rising.*

Virginia *Virginia accounts for about one-fifth of the municipal solid waste exported from the city, accommodating it at sites in Charles City, Maplewood, and Middle Peninsula, among others.*

Annual tonnage

- 0–49,000
- 50,000–99,000
- 100,000–499,000
- 500,000 +

Garbage

How Fresh Kills Worked

1. Engineers would stake out an area to become the site of a mound. The mound, which would be built at grade, had roads built into it and was constructed to account for washouts caused by heavy rain.

2. A hydraulic crane dug garbage out of the delivery barge and dumped it into a pen.

3. A front-end loader lifted material from the pen and dumped it in a "pay hauler"—a huge dump truck.

4. The pay hauler drove to the bank, the edge of the current dumping ground. It backed as far in as possible and dumped its load.

Fresh Kills Over a period of half a century, New York City relied on the Fresh Kills Landfill on the western shore of Staten Island. For years the world's largest sanitary landfill, it covers 2,200 acres—three times the size of Central Park—and claims mounds taller than the nearby Statue of Liberty (it was called a "sanitary" landfill because of the practice of covering each day's garbage with a layer of earth, as opposed to continuing to dump on top of open garbage). A network of marine transfer stations in four boroughs fed barges that were towed to Fresh Kills, carrying over 10,000 tons of residential waste each day.

Today, Fresh Kills consists of four mounds, or sections, which range in height from 90 feet to roughly 225 feet. Two of the four mounds are fully capped and closed: the remaining two are in the process of being closed. Numerous systems, previously put in place to protect public health and safety, continue to function. The gas recovery system forwards gas, a byproduct of decomposing garbage, either to the gas recovery plant located in one section or to flares located on the remaining three sections of the site. (Recovered gas is used for heating in roughly 10,000 Staten Island homes.) In addition, passive vents are located around the perimeter to ensure that no gas migrates off-site.

Although much of Fresh Kills today remains open waterway, intact wetlands, or wildlife habitat, years of work will be required to transform Fresh Kills into the park that is planned for the site. Planning alone is expected to cost over three million dollars, and to take several years. The site itself will take time to settle: it is estimated that natural settlement will reduce the height of the mounds by 10 to 15 percent over a 30-year period of time. Notwithstanding this extended time-frame, it is likely that some of the areas within Fresh Kills never used as landfill could open to the public as parkland as early as 2008.

From Dump to Park

1. The site contains 150 million tons of waste.

4. An impermeable liner was placed on the site.

Layers of a Landfill

Soil layer *At least six inches of planted topsoil cover active cells each day, with about a two-foot covering placed on the cell when it is full.*

Barrier protection layer *This layer helps direct rainwater run-off away from the landfill wherever possible and consists of concrete or gravel-lined drainage ditches.*

Compacted clay *Clay serves to reduce the permeability of the landfill.*

Geomembrane *Fabriclike mats may be used above or below the waste to protect the liner or further guard against rainfall infiltration.*

Gas venting layer *Gases produced by the decomposition of waste are either vented or collected and then burned off or processed.*

Solid waste

Leachate collection system *Drains and pipes collect contaminated water which is drained into a leachate pond.*

Naturally occurring soils at the site have a high clay content, which serves as a natural barrier to leachate.

5. A bulldozer pushed it in, and a compactor drove over it.

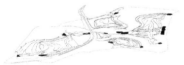

2. The initial task has involved liquid collection and containment.

3. Subsequently, a gas extraction network was put in place to collect methane produced by rotting garbage.

5. Drainage infrastructure will be placed to address storm water and roadways.

6. New habitats will be planted, and new pathways for users introduced.

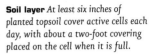

7. When completed, Fresh Kills Park will be among the largest parks in the city and will feature a variety of passive and active recreational offerings.

Garbage

Recycling   Over the last decade, the city's recycling program has become a fixture of daily life in New York. Initiated in 1993 as the largest curbside recycling program in the country, it involves collecting metal, glass, plastics, and paper from roughly three million households across the city. Although recycling volumes fell when glass and plastics collection were briefly suspended between 2002 and 2004, the program has returned to its previous levels of participation.

Today, roughly 19 percent of all material collected curbside is recycled—more than double the amount recycled when the program was introduced a decade ago. Paper is, not surprisingly, the biggest element of that total (paper represents roughly 40 percent of municipal solid waste nationwide). Five private companies have contracts with the city to buy its waste paper; one of them—Visy Paper—has a processing facility in Staten Island, which produces linerboard from the recycled paper.

Metal, glass, and plastic are collected separately from paper, and are currently delivered to one company—Hugo Neu—which barges the recyclables to its plant in Jersey City, where they are separated into individual materials and baled for export to processors. In a complicated commercial deal, the city gets paid by the company for the more valuable parts of the recyclables stream—the metal—while paying the company for disposing of the less marketable glass and plastic.

Until recently, the city's commitment to recycling was questionable: short-term contracts made it difficult for companies to invest in the automated materials-separation facilities that now exist in other places. But in 2004, the city announced the award of a 20-year contract to Hugo Neu. As part of that deal, the company agreed to develop a barge-served processing center on the waterfront in South Brooklyn.

Exporting Recyclables   Currently, sanitation vehicles deliver recyclables to Hugo Neu's sites in Hunts Point, Long Island City, and New Jersey. From the sites in New York, the recyclables are loaded onto a barge and transported across the harbor to the company's processing facility at Claremont Yard in Jersey City, New Jersey.

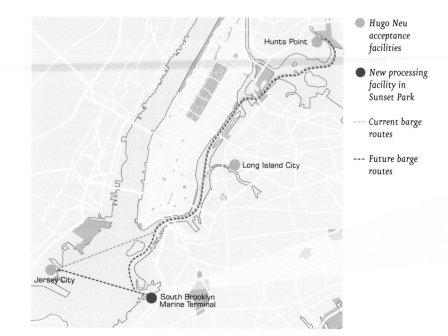

- Hugo Neu acceptance facilities
- New processing facility in Sunset Park
- --- Current barge routes
- --- Future barge routes

The Earliest Recycling

Recycling in New York City dates back 110 years, to the waning days of the nineteenth century. In 1895, as head of the Department of Street Cleaning, Colonel George Waring banned ocean dumping and instituted systematic recycling. Citizens were asked to separate ash, which was taken to landfills, and animal waste, which was sold to private contractors for fertilizer. The city also collected dry rubbish, such as rags and paper. What couldn't be recycled was burned in new municipal incinerators, generating enough electricity to run the plants. The ambitious recycling program was halted in 1918, on account of a shortage of labor and materials, and ocean dumping began again in earnest.

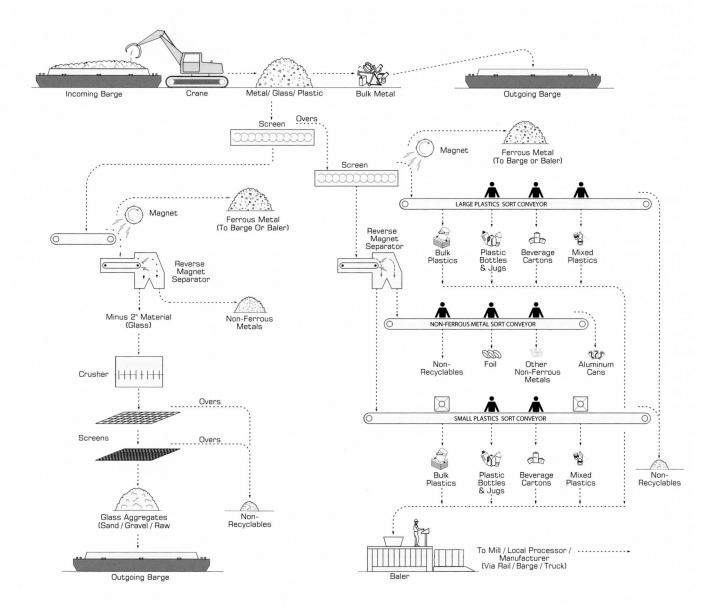

Materials Recovery, Hugo Neu Style

In the future, materials will be barged from these same
sites, plus one in Manhattan, to a new Materials Recovery
Facility (MRF) on 10 acres at the South Brooklyn Marine
Terminal in Sunset Park. There the materials will be source
separated, with the ferrous metals sent to domestic or
international steel users, the glass ground up and used for
aggregate, and the plastics turned into resins for reuse
in plastic products.

Incoming Barge Crane Metal/ Glass/ Plastic Bulk Metal Outgoing Barge

Screen Overs

Screen

Magnet Ferrous Metal
(To Barge or Baler)

LARGE PLASTICS SORT CONVEYOR

Magnet Ferrous Metal
(To Barge Or Baler)

Reverse
Magnet
Separator

Bulk
Plastics Plastic
Bottles
& Jugs Beverage
Cartons Mixed
Plastics

Reverse
Magnet
Separator

NON-FERROUS METAL SORT CONVEYOR

Minus 2" Material
(Glass) Non-Ferrous
Metals

Non-
Recyclables Foil Other
Non-Ferrous
Metals Aluminum
Cans

Crusher

SMALL PLASTICS SORT CONVEYOR

Overs

Screens Overs

Bulk
Plastics Plastic
Bottles
& Jugs Beverage
Cartons Mixed
Plastics Non-
Recyclables

Glass Aggregates
(Sand / Gravel / Raw Non-
Recyclables

Outgoing Barge Baler To Mill / Local Processor /
Manufacturer
(Via Rail / Barge / Truck)

Garbage

The Visy Paper Mill

on Staten Island recycles 1,200 tons of paper each day, producing linerboard that is used for corrugated packaging products. In the process, the company claims to use only about 10 percent of the electricity a virgin wood-pulping facility would use.

2. *A grappler crane picks up 12 to 14 tons at a time, and moves them to a pulper.*

3. *Water is added to the paper inside the pulper.*

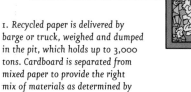

1. *Recycled paper is delivered by barge or truck, weighed and dumped in the pit, which holds up to 3,000 tons. Cardboard is separated from mixed paper to provide the right mix of materials as determined by a computer.*

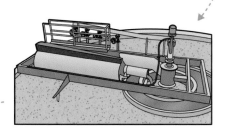

4. *The pulp is moved to a dump chest, and wax is removed. Good fibers are separated from the reject material, which is either composted or goes to a landfill. Chemicals and starch are added.*

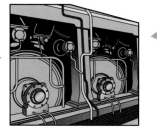

5. *Fibers are combined and go through about 30 felt presses or rollers. The paper is pressed and dried, and water is squeezed out.*

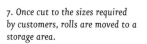

6. *Paper is rolled into jumbo rolls.*

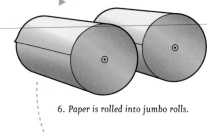

7. *Once cut to the sizes required by customers, rolls are moved to a storage area.*

8. *Rolls are later loaded onto truck trailers for delivery to customers.*

Composting In addition to paper, metals, glass, and plastics, the city recycles some organic materials separately from its regular household waste and recyclables. Roughly 20,000 tons of fall leaves are collected by the Sanitation Department each year; these are debagged and placed on unused city parks—one at Fresh Kills, another on the Brooklyn/Queens border, and a third in the Bronx. The leaves are formed into "windrows," or elongated piles, and left to turn to compost over a six- to eight-month period. The product is used as a fertilizer at various city sites.

The city also recycles Christmas trees, in a joint venture between the Parks Department and Sanitation. Sanitation alone collected over 100,000 trees last year—most of which were taken either to Fresh Kills, where they were composted, or to Randalls Island, where they were chipped into a mulch.

Food waste from city facilities is currently composted in only one location—Rikers Island, where an indoor facility on the island processes about 400 tons each month. The compost is used in a variety of horticultural projects across the island. A second on-site food waste composting project is being evaluated for Hunts Point food distribution center in the Bronx.

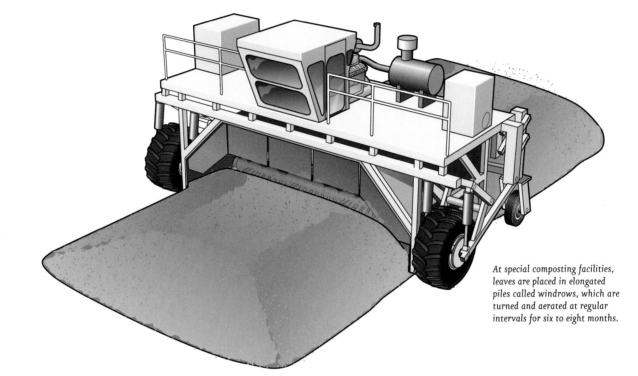

At special composting facilities, leaves are placed in elongated piles called windrows, which are turned and aerated at regular intervals for six to eight months.

Composting on Rikers Island The nation's largest municipal prison system, home to over 17,000 inmates and 7,000 correction officers, generates over 20 tons of food waste per day. Over 80 percent of this is composted at a facility on the site. The waste—which consists of vegetable trimmings, rotten food, leftovers, and the contents of scraped plates—is collected in specially marked 44-gallon plastic containers and then tipped into a Dumpster, which is transported to a compost facility on-site. At the composting facility, the contents of the Dumpster are emptied onto an open floor and mixed with wood chips, which supply carbon required to achieve rapid decomposition and also help aerate the waste. "Agitating" equipment blends the material and moves it forward through designated bays over a period of 14 days. By the time the material reaches the end of the bays, the food waste has been transformed into a soillike substance. This substance moves to an indoor curing area, where it stays for roughly a week, and then to an outdoor curing area, where it stays for about a month. After being screened to remove any contaminants, it is ready for use as a soil amendment or supplement.

Garbage

Commercial Waste

The term "commercial waste" generally refers to garbage produced by the city's businesses, including offices, retail stores, and restaurants. Each day, an estimated 13,000 tons of this kind of waste are produced. Much of this waste is paper, which is typically collected by private haulers from commercial buildings and delivered to recyclers, exporters, or paper manufacturers. The remainder, often referred to as "putrescible waste," is delivered to many of the same transfer stations now handling much of the city's residential waste—primarily in New Jersey, the Hunts Point area of the Bronx, the Greenpoint/Williamsburg area of Brooklyn, or in Jamaica, Queens. Both collections typically occur at night, when the streets are quietest.

In addition to commercial waste, New York produces over seven million tons of construction and demolition debris each year. Some 30 to 40 percent of it is one or another form of building waste (plasterboard, plumbing, etc.) and moves to transfer stations permitted to handle this kind of waste, many of them in the outer boroughs or New Jersey. The remainder is often referred to as "clean fill" and consists of gravel, dirt, rock, concrete, and stone that is largely the result of site excavation and building demolition. It is crushed and milled at a "clean fill" milling operation and used by both public and private sectors as soil cover in a variety of places.

Commercial Waste Generation
tons generated daily

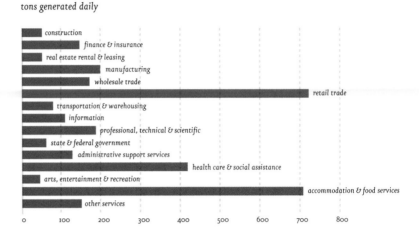

Waste and "the Mob"

Until the middle of the twentieth century, the city bore responsibility for collecting commercial waste as well as residential waste. That changed in 1957, when the city withdrew from the commercial waste business and the private carting industry—and organized crime—took over. Some private carters found themselves members of trade associations controlled by organized crime, and individual businesses found themselves with little choice of provider: specific locations were "owned" by specific haulers and the local business owner had little or no say in how much they paid for waste removal.

In the mid-1990s, after much effort on the part of law enforcement agencies, indictments were handed down against the leaders of the trash cartel in the city. At the same time a new local entity, the Trade Waste Commission, was set up to regulate the commercial waste haulers by licensing valid businesses and setting maximum and minimum rates for waste collection and disposal. Although its name has been changed to the Business Integrity Commission, the organization's oversight role continues today.

Roosevelt Island With one exception, New York City's garbage collection is pretty low-tech—trucks, cans, and a lot of manual labor. The exception is Roosevelt Island's "Automated Vacuum Assisted Collection" system—otherwise known as AVAC. Designed to handle refuse in densely populated areas, it relies on a series of chutes which run vertically through the island's high-rise buildings and are connected to a 20-inch pipe running under the island. Centrifugal turbines

pull the garbage at 60 mph through a vacuum to a central facility for collection, compaction, and containerization. Manned by nine Department of Sanitation employees, the system has the capacity to collect trash from 20,000 people, making it the largest of only 10 such systems operating in the country (DisneyWorld has one as well). Currently it processes eight tons each day, though expansion to nine new apartment buildings on the island is planned.

The AVAC in Action

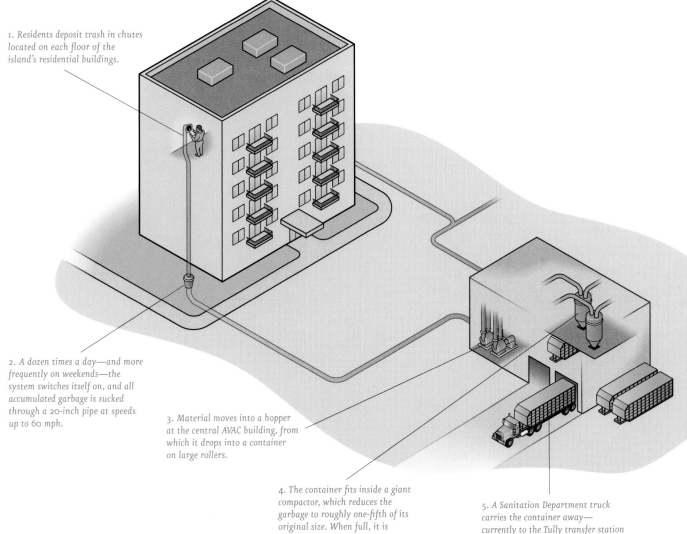

1. Residents deposit trash in chutes located on each floor of the island's residential buildings.

2. A dozen times a day—and more frequently on weekends—the system switches itself on, and all accumulated garbage is sucked through a 20-inch pipe at speeds up to 60 mph.

3. Material moves into a hopper at the central AVAC building, from which it drops into a container on large rollers.

4. The container fits inside a giant compactor, which reduces the garbage to roughly one-fifth of its original size. When full, it is rolled away for removal.

5. A Sanitation Department truck carries the container away—currently to the Tully transfer station in College Point, Queens—for disposal. Five or six of these large, compacted containers are removed from the island each week.

Garbage

Street Cleaning Most New Yorkers know little about street cleaning other than that it's one of the main reasons they can't leave their car parked on a street for days at a time. Few recognize how important a role the mechanical street sweepers and their drivers play in keeping city streets clean.

Each day, the Sanitation Department sends out roughly 325 sweepers. Though the machines travel slowly, generally covering between six and 20 miles a day, they are actually capable of moving much faster if necessary: thanks to a Mercedes-Benz engine, their top speed is a swift 37 miles per hour. Each sweeper holds 5.5 yards of debris, and will dump into collection trucks—generally twice a day, but more in leaf season.

The key to a successful sweep is water, and each of the sweeper vehicles holds 240 gallons of it. When they run out of water en route, drivers will refill their tanks at predetermined hydrants with special magnetic caps on them. Like the Fire Department, the street sweeper driver is guardian of a special wrench that gives him or her the ability to open the designated hydrants.

Underneath the Street Sweeper

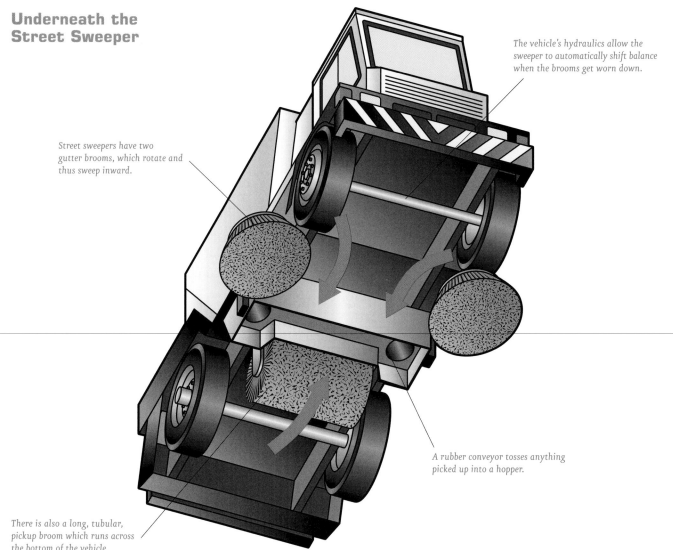

The vehicle's hydraulics allow the sweeper to automatically shift balance when the brooms get worn down.

Street sweepers have two gutter brooms, which rotate and thus sweep inward.

A rubber conveyor tosses anything picked up into a hopper.

There is also a long, tubular, pickup broom which runs across the bottom of the vehicle.

Policing the Streets

The Sanitation Police are a part of the department's enforcement division, whose white cars are not uncommon sights on New York City streets. The division's job is to ensure that both residents and businesses comply with health and sanitary laws governing both waste disposal, including recycling, and street cleanliness. The division also operates an Illegal Dumping Task Force and a bounty program to reward people who report illegal dumping.

How Clean Is Clean?

Street cleanliness is measured by the Mayor's Office of Operations, and ratings are based on rigorous photographic standards of cleanliness. Inspectors are trained to assess conditions based on a scale of I (cleanest) to 3 (dirtiest), with ratings below 1.5 considered "acceptably clean." Inspections may occur before or after street cleaning activities, and not all streets are visited: sample streets are statistically and geographically representative of a given district. The information produced is provided to Community Boards, Business Improvement Districts (BIDs), and other public interest groups interested in local conditions.

Street Cleanliness Ratings 1980–2003

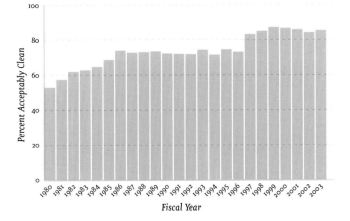

Fiscal Year

Acceptably clean: 1.0 *a clean street, with no litter*

Acceptably clean: 1.2 *a clean street, with just a few traces of litter*

Not acceptably clean: 1.5 *large gaps between pieces of litter*

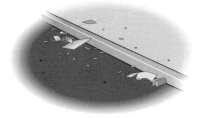

Filthy: 1.8 *litter is concentrated in spots, with gaps between piles or pieces of litter*

Filthy: 2.0 *litter is concentrated, with gaps between piles*

Filthy: 3.0 *litter is very concentrated, both in a straight line along and running over the curb*

Garbage

Snow Removal Although most of its day-to-day resources are devoted to garbage removal and street cleaning, the Department of Sanitation endears itself most to New Yorkers during snowstorms. With the first hint of snow, the department readies itself for action. Depending on the forecast, a certain number of sanitation trucks are "dressed"—fitted with plows in the front and chains on the back tires. Once two inches cover the ground (they can't plow less than that due to a mount on the vehicle which protects them from uneven manhole covers and other obstacles), they hit the streets.

To deal with particularly large snowstorms, the department relies on snow melters—diesel-powered machines that are towed into place by a tractor cab and can melt as many as 60 tons an hour. The machines are relatively new—a response to the late 1990s ban on dumping salty snow into the river—and have proved effective at tackling the larger mounds that can clog intersections and streets. Built with extra-large holding tanks, the machines melt the snow and direct the water runoff into nearby storm drains.

Let It Snow...

= 10 trucks

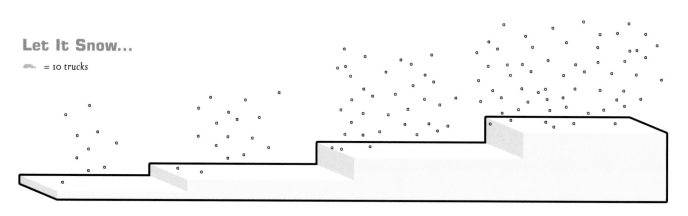

With only one inch of snow, only salt spreaders—350 of them, as well as 60 small pickup/salt spreaders—are used.

When there is a snowfall of between two and four inches, all spreaders are used and one-quarter of the Sanitation truck fleet—roughly 380 vehicles—is equipped with plows and sent out.

At four to six inches, between two-thirds and three-quarters of the fleet will be used—over 700 trucks.

A big snowfall, more than six inches, will involve equipping 100 percent of the fleet (1,335 trucks) with plows.

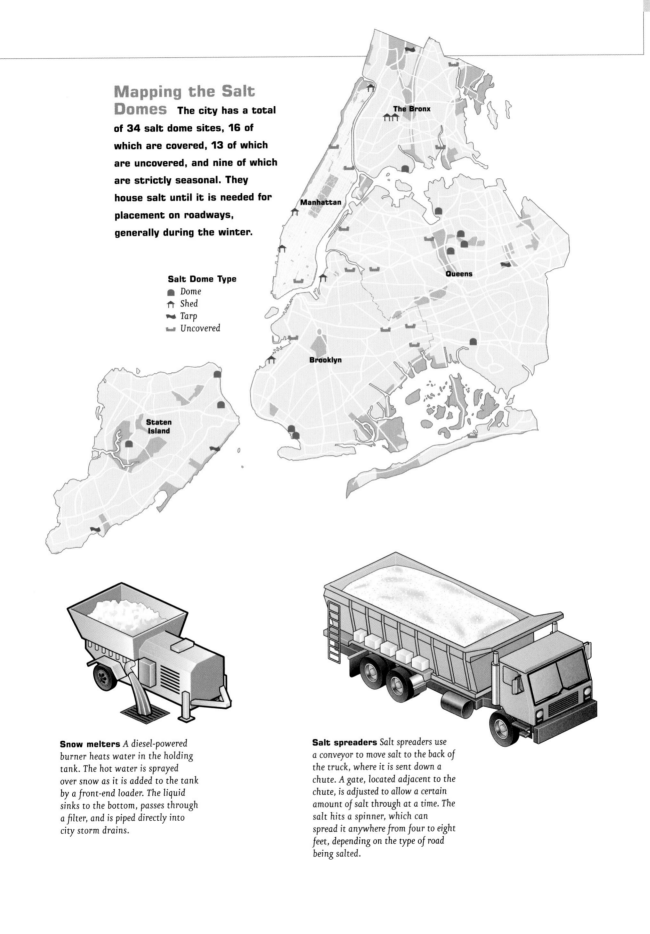

Mapping the Salt Domes

The city has a total of 34 salt dome sites, 16 of which are covered, 13 of which are uncovered, and nine of which are strictly seasonal. They house salt until it is needed for placement on roadways, generally during the winter.

Salt Dome Type
- Dome
- Shed
- Tarp
- Uncovered

The Bronx

Manhattan

Queens

Brooklyn

Staten Island

Snow melters *A diesel-powered burner heats water in the holding tank. The hot water is sprayed over snow as it is added to the tank by a front-end loader. The liquid sinks to the bottom, passes through a filter, and is piped directly into city storm drains.*

Salt spreaders *Salt spreaders use a conveyor to move salt to the back of the truck, where it is sent down a chute. A gate, located adjacent to the chute, is adjusted to allow a certain amount of salt through at a time. The salt hits a spinner, which can spread it anywhere from four to eight feet, depending on the type of road being salted.*

THE FUTURE

Looking back over the last century, it is fair to say that the layers of New York's infrastructure took form haphazardly in response to discrete civic needs and at different times—for streets and clean water in the nineteenth century, for efficient transport at the dawn of the twentieth, for speedier means of trade by the middle of that century. It is not surprising, then, that the systems

portrayed here operate almost entirely independently of one another and share little or nothing in the way of physical infrastructure.

As we look to the future, however, it is likely that connections between forms of infrastructure will become more routine—in New York and elsewhere. Telecom lines may well run through sewers, solar power will be used at sewage plants, wastewater can replace freshwater for industrial uses, and garbage may be among the fastest growing rail freight commodities. And as parts of New York are reimagined in their entirety—lower Manhattan and the West Side Yards above all—so too will be the infrastructure that supports them: green power, gray (reused) water, new subway lines and tunnels, and possibly the first expansion of the city's steam system in decades.

To the casual observer, these changes are relatively minor. A third water tunnel will facilitate local water distribution, but it only reinforces the city's commitment to its upstate watershed delivery system. New freight rail routes will open, but they will support—rather than supplant—vehicular tunnels and will rely almost exclusively on rail lines laid out generations ago. Channels will be dug and blasted deeper, but they represent almost precisely the maritime trade lanes that ships have plied for centuries.

Over the past two centuries, New York's infrastructure has successfully supported the growth of the city around it. The durability and longevity of the region's infrastructure is a testament to the vision of those who laid it down in some cases more than a century ago—who saw the need for express tracks on the subway, for open stretches of swamp for containers, and for upstate land for clean water. But it is also a tribute to the thousands of municipal and private sector workers who, for generations, have toiled around the clock to maintain the way of life that New Yorkers have come to expect.

Moving People

As we look to the future, it is hard to imagine traffic not moving faster on the region's roads: tollbooths and E-ZPass will become relics of the past, as electronic identification is built into individual cars; traffic warnings will become more precise and timely, as new ways to monitor data about road conditions become available; and low-tech improvements to mitigate traffic in the most congested areas of the city will continue to be made by city DOT.

But while roadway congestion may be the focus of daily life in Los Angeles, Dallas, or Miami, in New York it is rail transportation that—more than any other mode of travel—knits together people and their workplaces. And it is changes to that system in the future that will contribute most to more efficient movement of people in the future.

Three new initiatives are likely to improve regional rail connections. The first, the Second Avenue Subway, will relieve the severe overcrowding on the numbers 4, 5, and 6 lines that are the sole north-south subways on the East Side of Manhattan. The next, the proposed subway to JFK, would offer direct rail access to the airport as well as dramatically improve travel times for Long Island commuters who work in lower Manhattan. The third, a proposed new tunnel to increase direct rail service from New Jersey into midtown Manhattan, is likewise aimed at improving the commuter experience—in this case for those living west of the Hudson.

Second Avenue Subway

There are many projects on the drawing board in the headquarters of the Metropolitan Transportation Authority, but none awaited so eagerly—or for so long—as the Second Avenue Subway. One quick glance at a New York City subway map suggests why: while the West Side of Manhattan has multiple lines running south from the Bronx, the East Side has only one: the notoriously overcrowded Lexington Avenue line.

As early as the 1920s, when it was proposed as phase two of the plan for the new city-owned IND system, there was strong public support for the Second Avenue line. But plans for the project were shelved due to the tight fiscal environment following the Depression and did not reappear until after World War II.

In the early 1950s, $500 million in New York State bonds were issued to support construction of the line, but the proceeds were spent on more urgent capital projects required by the then-aging system. Plans for the line were revived with the establishment of the Metropolitan Transportation Authority in the late 1960s, and small portions of the subway were subsequently constructed in East Harlem and the Lower East Side, but the project once again fell victim to New York City's financial woes.

With the recent allocation of federal funding, the project has returned to life. The current plan is to build a new line along the length of Second Ave., from 125th St. to Hanover Square, although the speed at which this occurs will be determined by the availability of additional local funding.

Direct Rail to JFK

For almost two decades, the MTA operated the "train to the plane": an hour-long journey on the A train to the Howard Beach station in Queens, where a bus would wait to shuttle riders to their terminals. So light was traffic on the line that no one complained, and few even noticed, when the service was scrapped by the MTA in 1990.

The idea of a direct rail link to John F. Kennedy Airport from Manhattan was revived after September 11, 2001, more as a component of an economic development strategy for lower Manhattan than as a solution to any particular transportation problem. A new rail link between the airport and lower Manhattan would not only make the latter a more attractive office location for business travelers, but by connecting at Jamaica with the Long Island Railroad would facilitate the journeys of hundreds of thousands of Long Island commuters destined for lower Manhattan, who today must switch to a downtown-bound subway at Penn Station.

Forty alternative routes were evaluated during the course of 2004 by consultants hired by the Lower Manhattan Development Corporation, the Port Authority, the MTA, and New York City. Ultimately, a route that involves extending the existing Air Train line (which runs from JFK to Jamaica Station) along the Atlantic branch of the LIRR and through a new tunnel between downtown Brooklyn and lower Manhattan was recommended. Although the projected cost of the tunnel is considerable (approximately six billion dollars), and the public review process extensive, moves are afoot to secure a large share of federal monies for the project.

Access to the Region's Core (ARC)

A new passenger rail tunnel entering Manhattan from the east could well be matched by one from the west, if bistate transportation planners have their way. A comprehensive study of a new two-track tunnel under the Hudson River from Secaucus into Penn Station was recently completed under the supervision of the Port Authority and New Jersey Transit. The study suggests that completion of a new tunnel, along with new passenger facilities near Penn Station, could double west-of-Hudson rail capacity, speeding up travel times and providing nearly all commuters from New Jersey, as well as those from Orange and Rockland counties in New York, with a one-seat ride into midtown Manhattan.

The tunnel would operate much the same way the existing Penn Station tunnel operates today: diesel-electric locomotives would operate under diesel on the branch lines in the suburbs but switch to electric power to travel through the new tunnel into Penn Station. Although the precise alignment of the new line has not been determined, it is expected to run under the Palisades, very close to the existing Northeast Corridor track that carries Amtrak trains into Penn Station today.

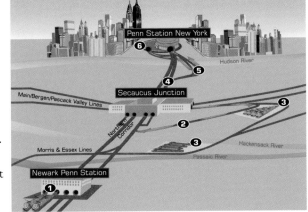

The estimated cost of the tunnel is a whopping three to five billion dollars.

Moving Freight

In some respects, the technology used to move goods—as opposed to people—has changed only modestly over the course of centuries. The bulk of international trade still moves in ships, much as it did at the time New York City was founded. The ships are bigger, faster, and more reliable— and much of the cargo is stuffed inside containers for ease of handling along the shore—but the mode of transportation is essentially the same. Somewhat more change has occurred with respect to the inland portion of the move, where trucks and trains have largely replaced stagecoaches and canals, but the overall pace of change has been slow—and is likely to remain that way for the foreseeable future.

There are nevertheless a handful of projects afoot in the metropolitan area that will, if they come to fruition, facilitate and potentially reconfigure local freight handling. Two of them—the Cross-Harbor Tunnel and the Staten Island Railroad—are rail projects, aimed at taking trucks off the congested local highways and expanding activity on the region's underutilized rail infrastructure. The tunnel is much more speculative, and much more costly, than the railroad—which is fully funded and likely to be completed in 2007.

The third freight project highlighted here involves the Bayonne Bridge. As container ships continue to increase in size, both harbor depths and bridge clearances will need to accommodate them. Considerable resources have been put into ensuring deep water in local shipping channels, but as of yet no formal commitments have been made to address the first of the harbor's bridge-clearance problems: the Bayonne Bridge.

Cross-Harbor Tunnel

The idea of a cross-harbor freight tunnel is not a new one; planners as far back as the turn of the last century saw a freight-only tunnel under the harbor as the ideal way to reduce congestion on waterways in the harbor. But with the opening of the Holland and Lincoln tunnels for vehicles, cargo found a new, speedier way to cross the river, and the idea of a rail freight tunnel was shelved.

Today, trucks crossing the Hudson continue to carry the bulk of the region's cargo: less than 2 percent of all freight enters areas east of the Hudson by rail. Rail freight coming to the region from the west and south must travel over a crossing at Selkirk, 140 miles to the north of the city, or make use of the limited carfloat service operating between Greenville Yards in New Jersey and Brooklyn.

Keen to reduce reliance on trucking and increase rail freight options, the federal government recently funded a major study on the impact of a freight rail tunnel between Brooklyn and New Jersey. The proposed tunnel—consisting of either one or two tubes—would connect the Greenville Yards in Jersey City to the Bay Ridge Line in Brooklyn, and terminate at an intermodal yard in Queens, from which the rail cargo would be distributed by truck. The cost of the proposed tunnel is high (between five and seven billion dollars) and local community opposition is strong, making it likely that the rail tunnel will continue to be the subject of discussion for many years to come.

Staten Island Railroad

New York is getting a new freight railroad. Abandoned by its previous operator a decade ago, the Staten Island Railroad—which spans the Arthur Kill and connects industrial businesses in Staten Island with the domestic rail system in New Jersey—is being brought back to life in a joint venture of the Port Authority and the City of New York.

Acquired by New York City from CSX Railroad, the railroad will provide direct rail service to the expanding New York Container Terminal at Howland Hook, to the Department of Sanitation's new containerization facility at Fresh Kills, and to industrial businesses on the railroad's Travis branch, which runs down the western shore of Staten Island. Reconstruction involves not only new track and replacement of rotting wooden-trestle creek bridges, but the completion of improvements to the Arthur Kill lift bridge, one of the region's few rail bridges over water. The bridge has been painted cobalt blue—the color of the original railroad serving that part of Staten Island.

Reactivation of the railroad will do more than meet the needs of Staten Island's freight-dependent businesses. It is expected to reduce truck traffic on the island by roughly 100,000 trips a year when it opens, in 2007.

Bayonne Bridge

The Bayonne Bridge, which connects Bayonne to the north shore of Staten Island, remains one of the world's longest steel arch bridges—its central arch spans 1,675 feet. Although it won an award soon after its opening in 1931 as the most beautiful steel arch bridge, today the bridge is something of a stepchild to the region's busier and  better known suspension bridges—the George Washington, the Brooklyn, and the Verrazano, among others. It has geography to thank for that: located in something of a metropolitan backwater, it handles only about 20,000 vehicles each day and is largely invisible to the great majority of New Yorkers.

But the bridge is anything but invisible to the pilots and captains of cargo vessels calling at the Port of New York and New Jersey. At 151 feet above sea level at midspan, the bridge is a navigational hazard to the larger container ships that make their way through the Kill van Kull off Staten Island en route to Port Newark/ Elizabeth. To date, several have collided with the bridge, forcing its temporary closure but not resulting in any serious structural damage.

To minimize conflicts in the future, the Port Authority of New York and New Jersey is evaluating the options for raising the bridge. The most likely plan involves leaving the bridge structure intact, but raising the roadway to provide greater clearance to ships. However, the estimated cost of such a plan could be several hundred million dollars, a sum unlikely to be paid back by fees from the relatively small number of cargo ships that stand to benefit from the removal of this rather unusual navigational hazard.

Power

Over the next several decades, the technologies used to produce electric power are likely to change relatively little.

However, as power plants get harder to locate within city boundaries, transmission projects bringing power generated outside of the region to the city are likely to become more attractive. So too are alternative energy sources, which should prove more competitive as the technologies mature and the cost of reliance on traditional fossil-fuel sources increases.

Two projects under consideration involve renewable energy sources. The first relies on water power, an important source of energy for the state as a whole but one that is largely absent, to date, from the city's energy mix. Another proposal focuses on wind power. Open stretches of ocean off the south coast of Long Island would be used to harness the wind's energy, delivering relatively modest amounts of power to the New York State grid or, possibly, to particular local facilities.

A third project addresses the region's need for natural gas. Existing pipelines often operate at capacity, and a variety of new proposals to expand pipeline capacity are making their way through the regulatory process. But rather than building a new pipeline, at least one company has proposed expanding the use of liquefied natural gas (LNG) to meet the region's growing need for gas. The proposal to moor an LNG delivery vessel in the middle of Long Island Sound, if successful, would mark a departure from the region's historic reluctance to locate infrastructure facilities offshore.

Hydro Turbines

The world's first bank of tide-powered turbines is likely to be developed here in New York City—in the middle of the East River. The plan is for six electricity turbines owned by Verdant Power, a Virginia-based energy company, to be attached to concrete piles driven into bedrock some 30 feet below the surface of the river. The heads of the turbines will move to face the current, and the blades will spin according to the ebb and flow of the tides, producing a modest 200 kilowatts of power— enough to provide energy to roughly 200 houses—at  their peak. Initially, the power produced will be delivered to two Con Edison customers on Roosevelt Island.

If the pilot is successful, however, the field of turbines could grow significantly. New York State has articulated a target of 25 percent reliance on renewable energy sources statewide by 2013, and tide power would join hydroelectric, wind, solar, and geothermal as means to that end. Verdant envisions a field of 200–500 turbines in the East River between Manhattan and Roosevelt Island.

Tide projects to date—most of them overseas—have generally acted more like dams, relying on barriers to hold back tidal waters to power generators. Only a few tidal mills have ever been put into use: the first 300-kilowatt turbine was installed off the coast of Devon in England in 2003 and another, of similar size, was located near Hammerfest, Norway.

The Long Island Offshore Wind Initiative

Long Island is the site of an ambitious proposal to harness the region's wind power. Known as the Long Island Offshore Wind Initiative, the plan—initiated by the Long Island Power Authority (LIPA) under a power purchase agreement with developer Florida Power and Light Energy—involves the placement of 40 wind turbines in an eight-square-mile area to the southwest of Robert Moses State Park, on the island's southern shore. If it moves ahead, the project could produce up to 140 megawatts each day as early as 2007—enough to power 42,000 homes in the service area.

Wind energy is the world's fastest-growing source of energy, in part because the costs of producing it have dropped roughly 80 percent in the last three decades. While the UK and continental Europe have relied on wind power much more extensively than the United States to date, states like California have successfully harnessed wind power since the 1970s and the federal government has articulated a goal of relying on wind power for 5 percent of the country's energy needs by 2020.

A successful offshore wind power project requires high winds, of course, but also relatively shallow water. Long Island's south shore is ideal in this regard, as it features shallow water (up to 70 feet is acceptable for wind turbines) as far as six miles out—meaning that the windmills, which extend up to 200 feet above sea level—might not be visible from the shore.

Liquefied Natural Gas

Nearly all of the natural gas used by the region's power plants, homes, and businesses is imported from the Gulf or western Canada via transcontinental pipeline. But reliance on the busy pipelines could be dramatically reduced if Broadwater Technologies—a joint venture of Royal Dutch Shell and TransCanada—succeeds in constructing and mooring a Floating Storage and Regasification Unit in the middle of Long Island Sound.

The shiplike terminal, roughly the size of the *Queen Mary 2* ocean liner, would receive liquefied natural gas shipments—natural gas "frozen" into a liquid state and kept at 260 degrees below zero Fahrenheit. At the terminal, the LNG would be warmed back into a gas and pumped into the existing Iroquois pipeline, which runs from Milford, Connecticut, to Northport on Long Island. The double-hulled vessel-cum-terminal would be able to store the equivalent of about eight billion cubic feet of natural gas. The proposed siting area is in the widest part of Long Island Sound, roughly 10 miles off the coasts of New Haven and Riverhead, in water 70–90 feet deep. From here, a 25-mile long line would run west along the seabed to meet the existing Iroquois pipeline.

Though the idea of an offshore facility is new, the use of LNG in New York is not. Both Con Edison and KeySpan maintain supplies of LNG locally, to meet peak demand during the coldest winter periods. When temperatures fall below 10 or 15 degrees Fahrenheit, LNG from on-site storage tanks is vaporized and sent out to meet system needs.

Communications

The world of communications, and telecommunications in particular, is evolving so rapidly that it is difficult to envision

how residents and businesses will communicate with each other two decades from now. It is much easier to identify a handful of projects that will expand either the accessibility or the reliability of telecommunications services throughout the city in the shorter-term future.

One project, placing wireless transponders on street pole tops, is actually part of a broader city initiative aimed at expanding wireless coverage throughout the city. In addition to pole tops, the city is actively identifying other property it owns or controls, such as schools, office buildings, and garages, that may be made available to wireless carriers that desire expanded coverage.

Another two projects are wholly private sector initiatives. One, broadband over power lines, is a technology that has been employed to a limited extent by utilities here and elsewhere for some time, but has yet to find a sizable commercial market. The concept is simple: take advantage of existing power lines as channels for information communication. The execution is more difficult.

In contrast, the other private sector project—the antenna proposed for the top of the new Freedom Tower in lower Manhattan—involves an old and proven technology: radio transmission. Once built, the project would simply involve the relocation of radio transmission activity from one place in the region—in most cases the Empire State Building—to another.

Wireless on Pole Tops

In an effort to improve mobile telephone services to city residents, New York City is planning to make available a variety of streetside poles under its control for transmission of mobile telecom services. This marks a departure from the evolution of the cellular market to date, which has largely relied on locating transmission equipment on private property—with no oversight by the Department of Information Technology and Telecommunications (DoITT).

To facilitate this process, DoITT issued a Request for Proposals in early 2004. Six companies were subsequently granted franchises, including one company that will be providing telephone access over the Internet—which should provide a lower cost alternative for phone service to low-income families. The franchises themselves run for a period of up to 15 years, and provide the company with the right to install and use telecommunications equipment on streetlights, traffic signals, and highway sign support poles.

Installation of the antennas is expected to begin in 2008. In addition to street-based poles, DoITT is also working with other agencies to identify other city facilities—including office buildings, schools, and garages—that may serve as useful transmission points to improve and expand cellular coverage across the city. A previous effort to locate cell-phone antennas on schools ran into significant community opposition and was withdrawn.

Broadband over Power Lines

The concept of using power lines for communication purposes is not new: utility companies have for years been using their own lines for various relay and control tasks. These applications require minimal bandwidth and low frequency, and have been largely successful. However, the use of power lines for high-bandwidth, two-way transmission involving retail customers has remained illusive—until now.

There are signs that the concept may soon come of age. Advances in chip design and electronics technology have improved delivery, and the explosion in demand for consumer broadband services has produced a stronger market. Equally important, deregulation in the electricity industry has taken firms like Con Edison out of the generation business entirely, leaving only their distribution and transmission assets as core businesses. Increasing the value of those assets by devising new revenue sources from them is more important than ever.

Only one utility, in Cincinnati, has rolled out broadband over power line technology on a commercial basis in the United States. Here in New York, a trial program is currently under way involving Con Edison and Ambient power line broadband systems: a number of electronic devices have been deployed at points within the power distribution system to overlay a separate communications network on the power lines. This network is delivering broadband services to a limited number of residences and a government facility. The trial has been successful to date, opening the door for commercialization of the technology at some point in the future.

Freedom Tower Broadcasting

The Freedom Tower, which will rise 1,776 feet above the ground at the former World Trade Center site, is intended to be the centerpiece of the rebuilding of lower Manhattan. It will also be the world's

tallest building, or at least the world's tallest freestanding structure, taking into account the broadcast antenna that will rise to 2,000 feet.

The antenna is designed to do more than simply pierce a height record. It is intended to be home to New York City broadcasters who historically relied on the antenna on the south tower of the World Trade Center. In specific, the Metropolitan Television Alliance—which includes city channels 2, 4, 5, 7, 9, 11, and 13—signed a Memorandum of Understanding (MOU) with Larry Silverstein, the Freedom Tower's developer, in 2003. The intention was to relocate their transmissions from their temporary home atop the Empire State Building.

The MOU marked a departure from the alliance's earlier plans—which was to build a 200-foot freestanding broadcast mast in Bayonne, New Jersey. Relying on the Freedom Tower would theoretically save money and effort, but it is not without its technical unknowns—most of them a function of the off-center location of the spire on the building's roof. Questions raised about the spire include how it might behave in high winds, what kind of material it should be made of so as not to interfere with broadcast signals, and whether the signal sent out from it would be blocked in any way by the building's shadow.

Keeping It Clean

The capital budgets of the Department of Environmental Protection and the Department of Sanitation are among the largest in city government, though much of what they build is not particularly new or interesting and is simply replacement for facilities that have grown too old or too decrepit to carry out their original mission effectively.

But there are now on the table a number of novel and challenging projects aimed at improving the health of the city. Two relate to improved management of the city's water supply. The Lloyd Aquifer project involves an experiment with "water banking" as a way to expand the city's ability to provide water to consumers in times of water shortage or drought.

The Croton filtration plant project is less notable for its proposed technology than for the fact that it is being built at all. Its opening will mark the first time that New York's water supply has been subject to the filtering requirements that have applied across the country for years, and will signal an end to a bitter battle over the siting of the facility in Van Cortlandt Park in the Bronx.

Another new initiative involves the city's garbage. For the first time since the announcement of the closure of Fresh Kills, the city is on the verge of embracing a new and relatively sophisticated system of waste disposal—one that relies on existing marine facilities to containerize the city's municipal waste and, in doing so, transform it from an environmental problem to a readily transportable export commodity.

The Lloyd Aquifer

For a number of years, the city's Department of Environmental Protection has been looking at ways of storing water underground to minimize the impact of droughts. Among the most promising technologies being explored is something known as "ARS"—aquifer recovery systems. The idea is to inject potable water from the upstate system into the Lloyd Aquifer—layers of porous rock deep beneath Brooklyn and Queens—during periods of surplus, and pump it out during times of need. The aquifer would serve as a sort of underground reservoir, able to provide water to a much larger audience than the barrier beach communities that it serves today.

Subterranean banking of water is a well-established practice in dry regions, such as California and Nevada, and in New Jersey—where the rock formations are not dissimilar to those existing in Brooklyn and Queens. Although it has never been used in the city, engineers are optimistic about its prospects: the clay cap of the Lloyd Aquifer would likely protect the water in the aquifer from sources of land contamination, and the freshwater would inhibit saltwater intrusion.

The ARS concept is expected to undergo a 14-month pilot test beginning in 2005: four observation wells will monitor the aquifer's performance, and injected water will be tested for research purposes before being discharged into the sewer system.

The Croton Filtration Plant

After a decade of battling, first with federal environmental agencies and then with local activists, the city is at last moving ahead with plans to build a water filtration plant under the Mosholu Golf Course in Van Cortlandt Park in the Bronx. This plant would filter the relatively small percentage—10 percent—of city water coming from the Croton system.

In most places across the country, federal law requires that drinking water be filtered to screen out impurities. While New York City has historically been granted a waiver from this mandate for its Catskill and Delaware water on account of their purity, the Croton system—located in a more developed and faster growing suburban area—does not meet the more stringent criteria for filtration avoidance. Since 1998, when the city was forced to enter into a consent decree with federal EPA requiring the completion of the plant on a certain timetable, it has been fined more than $400,000 for its failure to meet federal standards for drinking water.

The project is an expensive one, made more so by the stiff opposition the project aroused from neighborhood and environmental interests in the Bronx. To placate these interests, the city agreed to spend $220 million on unrelated improvements to parks in the Bronx—for playgrounds, running tracks, landscaping, equipment, and waterfront access—in addition to the estimated $1.2 billion cost of constructing the underground filtration plant itself.

Containerized Garbage

Like every county in the state, New York City is required to have a 20-year plan for disposing of its solid waste—but it doesn't. The relatively sudden closure of its historic landfill in Staten Island left it with a network of marine transfer stations and barges, but no destination for the waste. Since 2001, its marine transfer stations have lain dormant and most of its residential waste has been hauled to transfer stations in the outer boroughs, where it is shifted to larger tractor trailers for delivery to landfills in neighboring states.

All that will change if the city is successful in implementing its new 20-year plan for waste disposal. It envisions a network of reactivated marine transfer stations exporting containerized garbage by barge or rail to landfills in the Midwest and South. Containers of waste packed at the rebuilt transfer stations would be barged to either on-dock rail facilities in the harbor, where they would move to railcars for their journey to a disposal site, or to normal berths in the harbor, where they would be loaded onto oceangoing coastal barges.

The idea of containerized waste is not entirely new: a facility run by Waste Management operates at Harlem River Yards in the Bronx. But by embracing the concept of containerized waste so enthusiastically, the city is expecting to take an estimated three million truck miles off the road each year—a boon both to environmentalists and to residents in the communities now home to the transfer stations.

Acknowledgments There are many people who have contributed facts or information to this book, but only a handful responsible for what sets it apart from so many others. Having struggled for more than a year with book designers who couldn't fully grasp the idea of interweaving graphics and text, the partnership with Alexander Isley and his team has been more than serendipitous—it has been an honor and a pleasure. Their professionalism and enthusiasm over the past few months could not have been greater, and the book reflects both.

Among Alex's worthy crew, one name stands out above the others: George Kokkinidis. On the back of a one-day course with Edward Tufte, the dean of information graphics, George was able to conceptualize each and every section of this book— in ways that bordered on the spectacular. His ability to digest my conceptual vision for this book and translate it into a series of more than a hundred distinct and compelling spreads, while managing a far-flung team of graphic artists, has been an inspiration to us all.

What George has been to graphics, Wendy Marech has been to text—providing first-class research support over a period of two years. Although the research process may at times have been fun for her, it has been anything but easy. For all the people who cooperated in providing information or who took her on behind-the-scenes facility tours, there have been an equal number who refused to take her calls or answer her questions. Without her persistence and commitment—and there are very few cases in which it did not pay off—there would have been no book to write.

Myriad organizations and individuals provided information along the way. The Department of Environmental Protection; the Department of Information Technology and Telecommunications; the Department of Sanitation; the Department of Transportation; the Department of City Planning and the Economic Development Corporation—all of the City of New York—provided ongoing assistance. Among others, thanks are due to Marty Bellew, Mike Bellew, Tanessa Cabe, Agostino Cangemi, Alice Cheng, Tom Cocola, Ed Corbett, Walter Czwartacky, Rocco DiRico, Anthony Etergineoso, Magdi Farag, Renzo Ferrari, Salome Freud, Andrew Genn, Doug Greeley, Robert Kuhl, Venetia Lannon, Peter McKeon, Jeff Manzer, Ralph Mondella, Michael Mucci, David Nati, Henry Perahia, Gil Quiniones, Bruce Regal, Andrew Salkin, Jack Schmidt, Girish Shelat, Tom Simpson, Gerard Soffian, Harry Szarpanski, and John Tipaldo.

Other public agencies, including the Port Authority of New York and New Jersey, the New York State Department of Transportation; the U.S. Coast Guard, the MTA; the U.S. Army Corps of Engineers, the United States Postal Service, and Transcom also proved tremendously helpful. In particular, I am grateful for input provided by Bob Beard, Mike Bednarz, Amrah Cardoso, Jamie Cohen, Doug Currey, Bob Durando, Matthew Edelman, Bob Glantzberg, Lt. Michael Keane, Victoria Cross

Kelly, Don Lotz, Louis Menno, Ken Philmus, Joe Sardo, Joe Seebode, and Ken Spahn.

Many other individuals contributed their expertise, including Ken Stigner of Vollmer Associates, Bruce Lieberman of the New York & Atlantic Railroad, Sal Catucci of American Stevedoring, Captain W. W. Sherwood of the Sandy Hook Pilots Association, Matthew D'Arrigo of Hunts Point Cooperative Market, John Arrowsmith and Rich Wolf of ABC, Creighton Pritzlaff of Seagrave, Tom Peitz of FedEx, Ron Fridman of Duncan Parking Meters, Peter Scorziello of Synagro, Ellen Neises of Field Operations, Sam Schwartz of Sam Schwartz Engineering, Tom Schulze, Ted Olcott, David Lazecko, Jim Larsen, and the staff of the New-York Historical Society.

To those who helped get this idea out of the starting block—Silvio da Silva, Sloan Harris at ICM, and above all Ann Godoff at Penguin Press—I am most grateful. I am thankful, too, to those at Penguin who helped bring it over the finish line in what I consider record time, in particular Liza Darnton. But the largest debt of all is owed to those who patiently watched the marathon itself from the dining room sidelines—to my daughter Rebecca, who loves to write, and to my son Ned, who is passionate about how things work. For your good-natured acceptance of yet another adult activity in our busy home, I can say only the biggest of thank-yous—and hope that one day you get as much pleasure in reading *The Works* as I have had in writing it.

Contributing Artists I am also indebted to
a number of superb illustrators and photographers
who are responsible for the art that in many
ways forms the backbone of this book. They struggled
mightily to meet our ambitious deadlines and
ultimately performed well beyond our admittedly
high expectations.

Aaron Ashley 6, 9, 12, 18, 78, 84, 96, 100, 106, 108,
115, 117, 120, 127, 129, 130, 146, 160, 166, 167, 172, 173, 175,
175, 178, 193, 196

Michael Fornalski 23, 39, 40, 52, 55, 66, 70, 84, 94, 105,
126, 127, 138, 147, 177, 180, 187, 190, 197

Roger Garbow 11, 14, 36, 49, 183

Hyperakt Design Group 24, 47, 71, 76, 87, 88, 89, 93, 98,
102, 103, 109, 111, 113, 118, 137, 141, 143, 147, 148, 154, 158,
165, 168, 169, 171, 174, 177, 179, 183, 186, 188, 191, 194, 203

Jim Kopp 12, 18, 30, 54, 58, 62, 67, 77, 83, 96, 102, 104,
106, 114, 128, 129, 137, 159, 163, 164, 176, 186, 199, 201, 203

Jason Lee 5, 7, 9, 19, 20, 41, 43, 50, 64, 72, 73, 74, 85,
88, 95, 107, 125, 130, 140, 143, 149, 189, 202

Mgmt. Design 3, 5, 8, 15, 20, 27, 28, 29, 31, 37, 38, 42,
45, 59, 69, 72, 78, 81, 82, 95, 99, 133, 134, 147, 148, 155,
161, 162, 181, 182, 198, 201

Arthur Mount 19, 21, 25, 32, 155

Randy O'Rourke 45, 140, 143, 160, 179, 201, 209

Seymour Schachter 14, 32, 161

Chris Schappert 7, 13, 23, 54, 129, 132, 135, 138

Mark Schroeder 144, 145

Marty Smith 23, 33, 34, 60, 61, 74, 119, 200

Steve Stankiewicz 10, 15, 16, 17, 22, 42, 46, 48, 79, 105,
112, 115, 121, 129, 132, 156, 165, 167, 192

Image Credits Pages 2, 44 (right), 68, 86, 116, 124,
136, 153 (two images): Collection of The New-York
Historical Society (negative no. 42616, negative no.
77578d, negative no. 72507, negative no. 50756,
negative no. 77582d, negative no. 77577d, negative
no. 50739, accession no. X.47, no. 74563).
4 (left): Transcom.
9 (bottom), 10, 20 (evolution of a street sign), 47 (map):
New York City Department of Transportation, Division
of Traffic Planning, 2005. Permission granted by
the New York City Department of Transportation.
15 (bottom): © 1998, 2001, Fund for the City of New York.
22 (top): Courtesy of the New York City Department
of Design and Construction, illustration by Charles
Hearn, Jr.
22 (bottom): Courtesy of the New York City Department
of Design and Construction.
24 USDA Forest Service Northeastern Research Station
and Northeastern Area State and Private Forestry.
(Available from Open Accessible Space Information
System (OASIS), www.oasisnyc.net)
26 (two), 44 (left): Science, Industry & Business Library,
The New York Public Library, Astor, Lenox and Tilden
Foundations.
49 (screen shot), 54 (bottom), 77 (right): Courtesy of the
Port Authority of New York and New Jersey.
49 (bottom): The National City Company, drawing by
Cass Gilbert, 1926.
53 (right): Courtesy of the MTA Bridges and Tunnels
Special Archive.
60 (bottom): New York Regional Railroad, Jersey City,
New Jersey.

63 *(right)*: Joseph A. Tischner.

67 Photograph by Preston S. Johnson, collection of Wayne D. Hills.

71 Photograph by A. McGovern.

72 *(screen shot)*: Courtesy of the United States Coast Guard Activities, New York.

75 *(screen shot)*: © Great Lakes Dredge & Dock Co.

79 *(left)*, 98 *(bottom left)*: New York City Economic Development Corporation.

80 Cradle of Aviation Museum, Garden City, New York.

92 *(left)*: Edison underground system in 1883, Museum of the City of New York; *(right)*: Edison's large dynamo-electric generator, 1881, Museum of the City of New York.

98 *(top)*: Entergy Corporation.

98 *(bottom right)*, 113 *(right)*: KeySpan Corporation.

99 *(right)*: Used by permission of the New York Independent System Operator.

106 *(bottom left)*: The Blizzard of 1888, New Street looking toward Wall Street. Photograph: Brown Brothers. Museum of the City of New York.

110 Street lamp, standard type. Museum of the City of New York.

111 *(bottom)*: Energy Information Administration, Office of Oil & Gas.

118 *(right)*: Courtesy of Con Edison.

125 *(bottom)*: Courtesy of the Empire City Subway Company, Limited.

133 *(photograph)*: Kimberlee Hewitt.

142 Photography Collection, Miriam and Ira D. Wallach Division of Art, Prints and Photographs, The New York Public Library, Astor, Lenox and Tilden Foundations.

145 *(top)*: Courtesy of Panasonic Broadcasting.

147 *(left)*: New York City Department of Information Technology and Telecommunications.

152 Fifth Avenue, looking south from 42nd Street, c. 1880. Museum of the City of New York.

164 *(bottom)*: Local Union #147 L.I.U.N.A.

169 *(right)*, 179 *(bottom)*: Reproduced with permission of the New York City Department of Environmental Protection.

170 Photograph by C. Smith & K. Stigner, Vollmer Associates LLP.

185 Picture Collection, The Branch Libraries, The New York Public Library, Astor, Lenox and Tilden Foundations.

191 *(top)*: Waste Management of New York, LLC.

192 *(bottom)*: Field Operations.

195 Hugo Neu Corporation.

207 Design by NJ Transit/Systra Consulting, Inc.

210 © 2004 Verdant Power, LLC.

213 Lower Manhattan Development Corp.

215 Courtesy of the New York City Department of Sanitation. Greely and Hansen, LLC and Dattner Architects.

ABOUT THE AUTHOR

Kate Ascher received her M.Sc. and Ph.D. in government from the London School of Economics and her B.A. in political science from Brown University. Kate formerly held positions in corporate finance overseas and at the New York City Economic Development Corporation and Port Authority of New York and New Jersey. She is currently a senior executive with Vornado Realty Trust.

THE
END